T0262032

Understanding Meiosis

Understanding Meiosis

Edited by **Morgan Key**

New York

Published by Callisto Reference,
106 Park Avenue, Suite 200,
New York, NY 10016, USA
www.callistoreference.com

Understanding Meiosis
Edited by Morgan Key

International Standard Book Number: 978-1-63239-600-6 (Hardback)

Printed in the United States of America.

Contents

Preface

Meiosis refers primarily to the cell division for reproduction. Meiosis, the procedure of producing gametes in preparation for sexual reproduction, has long been a focal point of concentrated research. It has been researched at the cytological, hereditary, molecular and cellular stages. Researches in model systems have exposed universal essential mechanisms while parallel studies in various organisms have led to the discovery of variations in meiotic methods. This book primarily focuses on the molecular and comparative study of meiosis via model systems. It collects various strands of examination into this enthralling and demanding field of biology.

The information shared in this book is based on empirical researches made by veterans in this field of study. The elaborative information provided in this book will help the readers further their scope of knowledge leading to advancements in this field.

Finally, I would like to thank my fellow researchers who gave constructive feedback and my family members who supported me at every step of my research.

Editor

Molecular and Comparative Study of Meiosis in Model Systems

Facing the Correct Pole: The Challenge of Orienting Chromosomes for Meiotic Divisions

Karishma Collette and Györgyi Csankovszki
Department of Molecular, Cellular and Developmental Biology, University of Michigan,
USA

1. Introduction

Meiosis is a specialized cell division that directs a diploid germ cell to produce haploid gametes (reviewed in Schvarzstein et al, 2010; Sakuno and Watanabe, 2009). Meiosis differs from mitosis, in that one round of DNA replication is followed by two rounds of chromosome segregation. In the first round, meiosis I, homologous chromosomes separate (reductional division), and in the second round, meiosis II, sister chromatids separate (equational division). Meiosis II is similar to mitosis in that in both processes, replicated sister chromatids orient away from each other (are said to be bioriented), and will be separated at the metaphase to anaphase transition. By contrast, meiosis I, represents special challenges to the cell. In the first meiotic division, sister chromatids must face the same pole (are monooriented), and instead homologs, connected by the chiasma, are oriented away from each other toward opposite poles (Fig. 1). This chapter discusses our current understanding of how sister chromatid monoorientation and homolog biorientation are achieved during meiosis I.

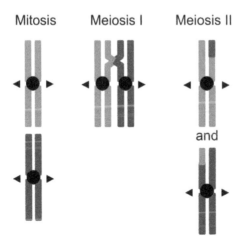

Fig. 1. Orientation of sister chromatids and homologous chromosomes in mitosis and meiosis. Arrowheads indicate the direction chromatids or chromosomes face.

Organisms with different chromosomal organization deal with this challenge differently. The majority of eukaryotic model organisms studied (such as yeast, flies and mammals) have monocentric chromosomes. Monocentric chromosomes have a single centromere, the region on the chromosomes where the kinetochore assembles. During mitosis and meiosis the kinetochores serve as site of attachment for spindle microtubules (Sakuno and Watanabe, 2009). In monocentric organisms, the centromere is where many meiotic events are coordinated, including orientation of chromosomes. Holocentric chromosomes (in organisms such as the nematode *C. elegans*) lack a localized centromere. On these chromosomes the kinetochore assembles along the entire length of the chromosomes during mitosis and forms cup-like structures encompassing each homolog during meiosis (Schvarzstein et al., 2010) (Fig. 2). Comparisons of the strategies used on chromosomes with such diverse organizational features highlight the common themes and the conserved molecular factors implementing these strategies.

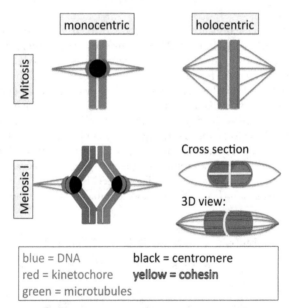

Fig. 2. The organization of kinetochores on monocentric and holocentric chromosomes during mitosis and meiosis I. During mitosis on monocentric chromosomes the kinetochore assembles at the region defined by the centromere. On holocentric chromosomes the kinetochores assemble along the entire length of the chromosomes. During meiosis I on monocentric chromosomes, kinetochores of sister chromatids function as one unit and orient toward the same pole. On holocentric chromosomes the kinetochore cups the volume of the sister chromatid pair.

One crucial player in the process of chromosome orientation is the chromosomal passenger complex (CPC), which monitors and regulates kinetochore-microtubules attachments. The CPC is composed of Aurora B kinase, inner centromere protein INCENP, Survivin and Borealin/Dasra B. Homologs of CPC subunits have been identified in organisms from yeast to humans. CPC proteins first associate with condensing chromatin at prophase, accumulate at the centromere during prometaphase and metaphase, relocate to the spindle midzone at the metaphase-anaphase transition, and finally associate with the midbody during telophase and cytokinesis (Vagnarelli and Earnshaw, 2004). This characteristically dynamic localization of the CPC likely reflects movement of Aurora B to act on different substrates. While on chromosomes, the CPC has important functions in condensin recruitment, facilitation of accurate microtubule-kinetochore attachments, spindle checkpoint function, and cohesin regulation. After the CPC transfers onto the central spindle and midbody, it is needed for the successful execution of cytokinesis (Vagnarelli and Earnshaw 2004).

The Structural Maintenance of Chromosomes (SMC) protein containing complexes condensin and cohesin, influence many aspects of chromosome organization and segregation and play roles in regulating chromosome orientation as well. Condensin is composed of a heterodimer of an SMC2 class and an SMC4 class ATPase protein, and three regulatory subunits referred to as chromosome associated polypeptide (CAP) proteins (Hirano, 2004; Hudson et al., 2009, Losada and Hirano, 2005). Budding and fission yeasts have a single condensin complex, but higher eukaryotes have two: condensin I and II. Condensins I and II differ in the identity of their CAP subunits. Condensin I contains CAP-G, -D2, and –H, while condensin II contains CAP-G2, -D3, and –H2. The single yeast condensin is most similar to the condensin I complex of higher eukaryotes. Although their exact molecular contribution to chromosome packaging remains a mystery, condensin complexes are essential for the precise organization and structural integrity of chromosomes (Bhat et al., 1996; Chan et al., 2004; Cobbe et al., 2006; Coelho et al., 2003; Dej et al., 2004; Gerlich et al., 2006; Hagstrom et al., 2002; Hartl et al., 2008; Hirota et al., 2004; Hudson et al., 2003; Lieb et al., 1998; Oliveira et al., 2003, 2005; Ono et al., 2003, 2004; Ribeiro et al., 2009; Samoshkin et al., 2009; Savvidou et al., 2005; Siddiqui et al., 2003; Stear and Roth, 2002; Steffensen et al., 2001; Vagnarelli et al., 2006; Watrin and Legagneux, 2005; Wignall et al., 2003; Yu and Koshland, 2003, 2005). SMC2 and 4 were originally identified as structural components of mitotic chromosomes in *Xenopus* and chicken cells (Hirano and Mitchison, 1994; Saitoh et al., 1994) and as important regulators of chromosome condensation and segregation in budding and fission yeast (Saka et al., 1994; Strunnikov et al., 1995).

Cohesin contains two SMC ATPase proteins of the SMC1 and SMC3 subclasses, and two regulatory subunits, Scc1 and Scc3 (Hirano, 2002, Jessberger, 2002). In meiosis, Scc1 is replaced by its paralog, Rec8 (Watanabe and Nurse, 1999). Cohesin mediates sister chromatid cohesion from the time they are replicated in S phase until sister chromatids separate at anaphase of mitosis and meiosis (Hirano, 2000; Nasmyth et al., 2001). Condensin and cohesin functions have been recently reviewed elsewhere. In this chapter we will concentrate on their role in regulating chromosome orientation.

2. Biorientation of sister chromatids in mitosis

Regulation of kinetochore orientation has been most extensively studied in mitosis. As the principle forces orientating chromosomes in mitosis and meiosis are likely mediated by the same factors, we will begin with a brief review of mitotic chromosome orientation. Mitotic chromosome segregation is preceded by a round of DNA replication, resulting in identical sister chromatids held together by cohesin. At entry into mitosis, or early during mitosis, condensin complexes associate with chromosomes to facilitate their compaction and segregation. In higher eukaryotes, the bulk of cohesin is removed from chromosomes arms in prophase, while centromeric cohesin is released at the metaphase to anaphase transition. In yeast, cohesin is removed from chromosomes in a single step, at the metaphase to anaphase transition (Nasmyth, 2002).

Prior to their separation, chromosomes align on the metaphase plate. Chromosome alignment and orientation is complete when all kinetochores are under tension, as a result of two opposing forces. Tension is generated because kinetochore microtubules are pulling sister chromatids toward opposite poles, and this poleward force is opposed by cohesion between sister chromatids in the centromeric region (Hauf and Watanabe, 2004; Sakuno and Watanabe, 2009). Sister kinetochores are arranged back-to-back, an arrangement that naturally facilitates biorientation. Kinetochore–microtubule attachments occur via trial and error, and are stabilized only when tension is established. An ideal attachment in mitosis, with two sister kinetochores attached to microtubules from opposite poles, is called an amphitelic attachment. Erroneous chromosome-spindle attachments, such as attachment of both sister kinetochores to the same pole (syntelic attachment), or attachment of only one sister kinetochore to a pole (monotelic attachment), result in insufficient tension. Merotelic attachment (one sister attached to both poles) could, in principle, establish tension, but must also be corrected (Fig. 3) (Cimini, 2007).

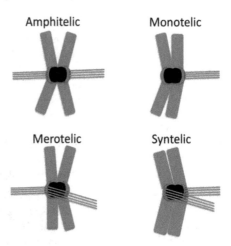

Fig. 3. Various forms of kinetochore-microtubule attachments at mitosis. Chromosomes are shown in blue, centromeres in black, kinetochores in red, and microtubules in green.

During mitosis, the CPC is enriched at the inner centromere where it is perfectly situated to monitor and correct aberrant kinetochore-microtubule attachments. (Tanaka, 2002; Vagnarelli and Earnshaw, 2004; Ruchaud et al. 2007; Watanabe, 2010) CPC depletion drastically disrupts metaphase chromosome alignment and orientation and subsequent anaphase segregation (Biggins et al., 1999; Biggins and Murray, 2001; Carvalho et al., 2003; Cimini et al., 2006; Ditchfield et al., 2003; Hauf et al., 2003, 2007; Honda et al., 2003; Kallio et al., 2002; Knowlton et al., 2006; Lens et al., 2003; Liu et al., 2009; Tanaka et al., 2002; Zhu et al., 2005). Aurora B and the CPC detect inappropriate or reduced tension at sites of incorrect attachment, and facilitate the release of these ill-fated connections. Aurora B appears to assume a more emphatic role in regulating syntely (Hauf, 2003, 2007; Lampson, 2004), but merotely is also fixed by the kinase (Cimini et al., 2006; Hauf et al., 2007; Knowlton et al., 2006). Hesperadin (a specific Aurora B inhibitor) treated HeLa cells showed a high incidence of syntelic attachments, and monotelic attachments were reduced six fold compared to control. These results led to a model whereby kinase function is required for the conversion of syntelic into temporary monotelic connections, such that the newly released kinetochore can search for microtubules from the correct pole (Hauf et al., 2003). A study in budding yeast suggests that centromeric INCENP-survivin (another CPC subunit) is a tension sensor that communicates with Aurora B to specifically destroy attachments under aberrant tension (Sandall et al., 2006).

Recent research has suggested that the spatial distribution of Aurora B is key to coupling tension sensing with the stability of kinetochore-microtubule attachments. Aurora B kinase influences microtubule interactions via phosphorylation of various centromeric/kinetochore targets. The phosphorylation status of Aurora B substrates depends on the extent of spatial separation from Aurora B at the inner centromere. Targets experiencing low tension (for example, in a syntelic connection) are physically closer to the kinase at the inner centromere, and undergo phosphorylation, whereas targets subjected to high tension (as seen in a bioriented attachment) are pulled away from Aurora B and escape phosphorylation (Tanaka et al., 2002; Liu et al., 2009).

Aurora B targets include the centromere associated kinesin MCAK (Andrews et al., 2004; Lan et al., 2004) and kinetochore proteins Dam1p, Ndc80p (Cheeseman et al., 2002) and Ndc10p (Biggins et al., 1999). Aurora B activation abolishes incorrect kinetochore-microtubule attachments by promoting microtubule fiber turnover and disassembly (Lampson et al., 2004; Cimini et al., 2006). Paradoxically, the microtubule-depolymerizing activity of MCAK, a major regulator of microtubule dynamics, is suppressed by Aurora B phosphorylation (Andrews et al., 2004; Lan et al., 2004). Perhaps phosphorylation of MCAK temporarily turns off its depolymerizing activity and supports new attempts at establishing kinetochore–microtubule connections, once incorrect attachments have been resolved (Knowlton et al., 2006; Lan et al., 2004).

Phosphorylation by Aurora B reduces the microtubule binding affinity of its kinetochore substrates (Cheeseman et al., 2006; DeLuca et al., 2006; Ciferri et al., 2008). Phosphorylation by Aurora B also disrupts subunit interactions within the multiprotein kinetochore and may thus contribute to the elimination of inappropriate kinetochore-microtubule connections (Shang et al., 2003). Thus, phosphorylation of the kinetochore targets of aurora B reduces their affinity for microtubules, causing them to relinquish wrong connections, and seek out fibers coming from the appropriate pole.

The CPC functions not only to detect and dissolve improper attachments, but it also activates the spindle attachment checkpoint (SAC) (Biggins and Murray, 2001; Carvalho et la., 2003; Ditchfield et al., 2003; Lens et al., 2003). SAC is activated in response to insufficient tension or unattached kinetochores. Checkpoint activation delays anaphase onset, so that bipolar attachments can be achieved for all chromosomes.

Tension is only established if the pulling forces of the spindle are counteracted by cohesion between sister centromeres. Centromeric cohesin function is needed for bipolar attachment of sister kinetochores (Sonoda et al., 2001; Tanaka et al., 2000). In fission yeast, cohesin localizes specifically to pericentric regions, and less to the core centromere (Bernard et al.,

2001; Nonaka et al., 2002; Tomonaga et al., 2000). These results support a model in which reduced levels of cohesin at the core centromere allow core regions to open up and assume back-to-back orientation, while pericentric cohesion ensures proper establishment of tension (Sakuno et al., 2009; Sakuno and Watanabe, 2009).

In addition to cohesin, condensin also seems to play a role in ensuring proper biorientation of sisters. In *C. elegans,* condensin depletion results in disorganized centromeres, and merotelic attachment (Hagstrom et al., 2002; Stear and Roth, 2002). In condensin depleted vertebrate cells, sister kinetochores move closer to each other and they do not face opposite poles anymore (Ono et al., 2004). It is likely that if the underlying chromosome structure is disrupted, especially at the centromeric regions, kinetochore function is perturbed as well.

3. Chromosomes are restructured extensively in preparation for meiosis

During meiosis I, sister chromatids orient toward the same pole and paired homologs (called bivalents) held together by the chiasma, become bioriented at metaphase instead (See Fig. 1). Differences in chromosome orientation during mitosis and meiosis could, in principle, be due to differences in the structure of the chromosomes themselves, or differences in the spindle. Micromanipulation experiments support the former possibility (Paliulis and Nicklas, 2000). Meiotic chromosomes placed on a mitotic spindle orient like meiotic chromosomes, and vice versa.

In both monocentric and holocentric organisms, chromosomes are extensively restructured during prophase of meiosis I to facilitate their proper segregation (reviewed in (Page and Hawley, 2003; Schvarzstein et al., 2010). Homologous chromosomes first pair, synapse (form a structure called the synaptonemal complex (SC) along the entire length of the homolog interface), and undergo crossing over, leading to the formation of bivalents. Metaphase chromosome alignment faces two challenges. First, sister kinetochores must be held together and orient toward the same pole. Second, homologs must be oriented away from each. As in mitosis, tension must be established. This tension is mediated by the pulling forces of the spindle microtubules exerted on kinetochores of homologs. Unlike in mitosis, the pulling forces of spindle microtubules are not counterbalanced by cohesion between sister kinetochores, but by the physical linkage of homologs mediated by the chiasma (see Fig. 2). Although research thus far has primarily focused on the importance of the CPC for sister chromatid biorientation in mitosis, recent evidence suggests that its role in destabilizing improper kinetochore-microtubule attachments may be conserved in meiosis as well (see below).

To deal with the special challenges of meiosis, the distribution and/or composition of chromosomal proteins can be different on meiotic chromosomes, as compared to mitosis. In mouse oocytes at metaphase I, condensin II concentrates within the core of individual sister chromatids, while condensin I is enriched at the centromere (Lee et al., 2011). This is in contrast to the mitotic distribution of these complexes. During mitosis in HeLa cells, the two condensin complexes alternate along chromosome axes, with condensin II being enriched at the centromere (Ono et al., 2004) (See Fig. 4). An additional difference between mitosis and meiosis is the subunit composition of cohesin. In meiosis, Rec8 (and other Rec8 paralogs) replace the mitotic paralog Scc1 (Watanabe and Nurse, 1999).

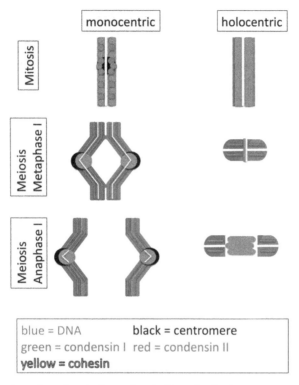

Fig. 4. The distribution of condensin I, condensin II, and cohesin on monocentric and holocentric chromosomes during mitosis and meiosis I.

In monocentric organisms, during prophase I of meiosis, cohesin molecules surrounding the site of crossover hold homologous chromosomes together until initiation of anaphase I. Homolog separation in meiosis I requires the resolution of sister chromatid cohesion distal to the chiasma. At the same time, sister chromatids must be held together until anaphase II by sister chromatid cohesion at the centromere. To achieve sequential loss of cohesin, in meiosis I arm cohesion is released, but centric cohesion is protected from degradation by Shugoshin/MEI-S322 (Kitajima et al., 2004; Resnick et al, 2006; Sakuno and Watanabe, 2009). This ensures that homologous chromosomes are no longer tethered, and can migrate to opposite poles, but sister chromatids travel together (Fig. 4) At anaphase II, centromeric cohesion is lost, and sister chromatids separate (Petronczki et al., 2003).

Holocentric organisms do not have a localized centromere, and phrases such as arm cohesion and centromeric cohesion do not apply. During *C. elegans* meiosis, the kinetochores form cup-like structures surrounding the entire volume of each homolog (Dumont et al., 2010) (See Fig. 2). Differentiation of domains within bivalents must therefore be coordinated in a different manner. Research in the last decade revealed that the formation of the crossover provides the primary clue for bivalent differentiation (Chan et al., 2004; Nabeshima et al., 2005). In *C. elegans*, each pair of homologs typically undergoes one crossover event in the terminal one-third of the chromosome. Toward the end of prophase I, bivalents are restructured into cross-shaped structures, in which the short arm is formed from the domain between the crossover and the closer chromosome end, and the long arm is formed from the domain between the crossover and the more distant chromosome end. Cohesin along the short arm is released in meiosis I to separate homologs, while cohesin along the long arm must be protected until anaphase II (reviewed in (Schvarzstein et al., 2010). Cohesion protection in C. elegans appears to be independent of Shugoshin, and instead is regulated by the HORMA domain containing protein HTP-1 and worm specific PP1 phosphatase interacting protein LAB-1 (de Carvalho et al., 2008; Martinez-Perez et al,. 2008). Note that while cohesion protection in meiosis I in monocentric organisms is performed at the centromere, in C. elegans cohesion is protected along the long arms of bivalents (Fig. 4).

As in monocentric organisms, distribution of chromosomal proteins differs between mitosis and meiosis in holocentric organisms as well. Our laboratory recently described the distinct distribution of condensin complexes in C. elegans, and compared their distributions in mitosis and meiosis (Collette et al., 2011). As in mouse oocytes, C. elegans condensin II occupies the core domain within each sister chromatid (Chan et al., 2004). Condensin I, on the other hand, is found at the midbivalent at the interface between homologs (Csankovszki et al., 2009; Collette et al., 2011) (Fig. 4).

4. Holding sister together: Monopolin and Rec8

Sisters are usually monooriented in meiosis I, even if homologs are not connected due to lack of chiasma formation. In most genetic backgrounds with defective crossover formation, homologs are not held together, but sister chromatids still segregate together toward the same pole (or are lost) during the first meiotic division. This finding implies that holding sisters together is not simply a consequence of keeping homologs apart. In budding yeast, a pair of sister kinetochores appears to be captured by a single microtubule (Winey et al., 2005). In higher eukaryotes, multiple microtubule attachments are made and both sister kinetochores are captured but they are connected to the same pole (Goldstein, 1981; Lee et al., 2000; Moore and Orr-Weaver, 1998; Parra et al., 2004.) Two possible mechanisms have been suggested to explain these findings: either sister kinetochores are fused into a single entity or one of the sisters is inactivated (Monje-Casas et al., 2007). Indeed, electron microscopy studies indicate that in meiosis I sister kinetochores are arranged side-by-side, as opposed to the back-to-back orientation seen in mitosis or meiosis II (Goldstein, 1981; Lee et al., 2000; Parra et al., 2004).

Studies of sister chromatid monoorientation in different model organisms have uncovered some interesting similarities and differences. Budding yeast makes use of a complex called monopolin to monoorient sisters. Monopolin is made up of Mam1, Csm1, Lrs4 and Hrr25 (Petronczki et al., 2003, 2006; Rabitsch et al., 2003; Toth et al., 2000). While in other model organisms the meiosis specific cohesin subunit Rec8 plays an important role in the process (see below), in budding yeast the function of Rec8 is not required for monoorientation (Toth et al., 2000, Yokobayashi et al., 2003), and monopolin appears to glue sisters together independent of cohesin function (Monje-Casas et al., 2007) On the other hand, condensin function is important for proper chromosome orientation, perhaps via correct localization of monopolin. In condensin depleted cells, monopolin association with kinetochores is reduced and a portion of sister kinetochores biorient (Brito et al., 2010).

Monocentric organisms other than budding yeast do not use Monopolin for sister chromatid monoorienation, and instead make use of cohesin and other meiosis specific factors. Rec8 seems to play an especially important role. Rec8 deficiency in fission yeast and plants leads to biorientation of sister chromatids, resulting in equational rather than reductional division in meiosis I (Chelysheva et al., 2005; d'Erfurth et al., 2009; Watanabe and Nurse, 1999, Yu and Dawe, 2000). Replacement of Rec8 with its mitotic paralog also results in biorientation, indicating that Rec8 function is specifically required (Yokobayashi et al., 2003). While Rec8 is necessary for sister monoorientation, it is not sufficient (Watanabe and Nurse, 1999; Yokobayashi et al., 2003). In fission yeast at least one other meiosis specific factor, Moa1, assists Rec8 in the process (Yokobayashi and Watanabe, 2005). In fission yeast, Rec8 has a distinct localization pattern, and is also found at the central core region of the centromere, in addition to chromosome arms and pericentric regions (Yokobayashi et al., 2003). This is different from the localization pattern of mitotic cohesin, which is specifically reduced at the centromere core (see above). According to current models, cohesion at the core centromere induces sister kinetochore monorientation, and cohesion at pericentric regions (as in mitosis and meiosis II) allows kinetochores to move away from each other promoting biorientation (Sakuno et al., 2009; Sakuno and Watanabe, 2009).

Since holocentric organisms do not have a localized centromere, they must use a different method to hold sister chromatids together during meiosis I. Recent studies suggest that despite the difference in chromosomal organization, REC-8 plays an important role in sister monoorientation in *C. elegans* as well (Severson et al., 2009). In *rec-8* mutant oocytes, homologs are not held together by a chiasma at metaphase due to an earlier defect in meiosis. However, *rec-8* mutant univalents (a pair of sister chromatids) become bioriented and separate from each other in meiosis I. Therefore *C. elegans* uses the same factor as monocentric organisms (REC-8), but this protein is functioning to promote sister monoorientation at different chromosomal domains: at the centromere in monocentric organisms and at the long arm of bivalents in *C. elegans*.

5. Regulating kinetochore-microtubule attachments: Aurora B function promotes homolog biorientation

As discussed above, in meiosis I homologous chromosomes become bioriented at the metaphase plate, as opposed to the biorientation of sister chromatids seen in mitosis and meiosis II. In principle, the problem to be solved is similar: tension must be established between two kinetochore entities that are connected. The difference is that this connection is centric cohesin in mitosis and chiasma between homologs during meiosis. During mitosis, Aurora B plays a crucial role in biorienting sister chromatids. During meiosis in fission yeast, budding yeast and mouse, Aurora B also localizes to the centromeric regions (Monje Casas et al., 2007; Parra et al., 2003, Petersen et al., 2001), therefore the protein is present at the right place and at the right time to regulate kinetochore-microtubule attachments in meiosis as well.

The first indication that Aurora B homologs are functioning in meiotic chromosome orientation came from studies in yeast. In budding yeast, Ipl1/Aurora B is needed for biorientation of homologs (Monje-Casas et al., 2007, Yu and Koshland, 2007). The results obtained in this system are consistent with the primary role of Aurora B to regulate turnover of kinetochore-microtubule attachments, similar to its role in mitosis. The fission yeast Aurora B homolog, Ark1, is also crucial for homolog biorientation during the first meiotic division (Hauf et al., 2007). Ark1 can promote biorientation of homologs, but it is also needed for biorientation of Rec8-deficient univalents (Hauf et al., 2007). In fact, Aurora B can promote biorientation even on kinetochores that are only loosely connected by a DNA thread (Dewar et al., 2004). These findings led to the suggestion that Aurora B can promote biorientation of any two kinetochores that are connected, further supporting the hypothesis that the molecular function of Aurora B is the same in mitosis and meiosis: correcting faulty kinetochore-microtubule attachments (Hauf et al., 2007).

Fission yeast Ark1 kinase appears to promote not only homolog biorientation, but also sister monoorientation. Ark1 can correct attachment of a unified pair of sister chromatids to both poles (merotelic attachment). While corrections of these merotelic attachments ultimately promote sister monoorientation, this mechanism is fundamentally different from the mechanism used by Rec8 and Moa1 to alter kinetochore geometry discussed above (Hauf et al., 2007).

On the holocentric chromosomes of *C. elegans*, Aurora B/AIR-2 is also located in an ideal position for regulating chromosome orientation, although this role has not been formally demonstrated yet. When bivalents orient at the metaphase plate, their long arms are parallel to spindle microtubules, and the midbivalent domains line up at the metaphase plate. Surprisingly, microtubule density on the poleward end of bivalents is low. Instead chromosomes are ensheathed by lateral microtubule bundles (Wignall and Willeneuve, 2009) (See Fig. 4). This ensheathment by lateral microtubules naturally promotes the orientation of homologs away from each other (Schwarzstein et al., 2010). Importantly, Aurora B/AIR-2 is found at the midbivalent domain, at the homolog interface (Kaitna et al., 2002; Rogers et al., 2002). Aurora B is perfectly situated here to promote homolog biorientation by performing its usual function: destabilizing incorrectly attached microtubules. Ensuring that microtubules do not attach at the zone of high Aurora B activity, coupled with ensheathment of bivalents by lateral microtubules provides an attractive model for biorientation of holocentric homologs.

6. Regulating chromosome orientation: a role for condensin?

Condensin I during mouse (Lee et al., 2011) and *C. elegans* meiosis (Collette et al., 2011) colocalizes with Aurora B. In the mouse, both condensin I and Aurora B are enriched at the centromere during metaphase I—at the domain important for regulating chromosome orientation. We recently demonstrated that in *C. elegans* AIR-2/Aurora B and condensin I colocalize at metaphase and anaphase of both meiotic divisions. Interestingly, condensin I occupies the same domain as AIR-2 not just on chromosomes at metaphase I and II, but also on the spindle between separating chromosomes at anaphase (Fig. 5). In addition, we also showed that correct targeting of condensin I to the midbivalent in meiosis I is dependent on AIR-2 (Collette et al., 2011). These results raise the interesting possibility that condensin I may aid Aurora B in performing its functions in regulating chromosome orientation in meiosis.

Metaphase I   Anaphase I

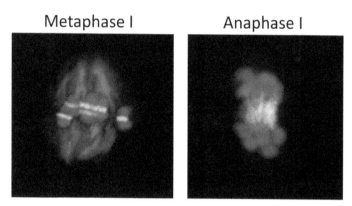

Fig. 5. Fluorescence microscopy images of condensin I distribution during *C. elegans* meiosis. During metaphase I, condensin I associates with the midbivalent domain, and during anaphase it is found on spindle microtubules between separating chromosomes. Microtubules are shown in red, condensin I in green, and chromosomes in blue.

Condensin depletion experiments support a role for condensin in regulating chromosome orientation. In budding yeast, condensin depletion leads to reduced localization of Monopolin subunit Mam1 to kinetochores and increased incidence of sister chromatid biorientation (Brito et al., 2010). In mouse oocytes, injections of SMC2 antibodies (which interfere with the function of both condensin complexes) led to a portion of sister chromatids assuming back-to-back, rather than side-by-side orientation. Anti–CAP-H injection (which interferes with condensin I only) caused disorganized centromeres and chromosome alignment problems (Lee et al., 2011). Condensin function may influence kinetochore structure, or may affect the underlying structure of centromeric chromatin (Bernad et al., 2011; Lee et al., 2011; Ono et al., 2004). Whether similar defects will be seen upon condensin I depletion in holocentric organisms as well remains to be demonstrated. However, given the colocalization of Aurora B and condensin I in mouse and *C. elegans* meiosis, the role for Aurora B in chromosome orientation in yeast and mouse meiosis, and the apparently conserved function of Aurora B in different organisms, we would like to propose that condensin I may aid chromosome orientation in *C. elegans* meiosis as well.

7. Conclusion

During meiosis in monocentric organisms, the centromere is used as a site where kinetochore-microtubule attachments, and therefore chromosome orientation, are regulated. In organisms other than budding yeast, enrichment of meiotic Rec8-containing cohesin at core centromeric regions promote monoorientation of sister chromatids. Aurora B at centromeric region appears to function to correct aberrant kinetochore-microtubule attachments to promote biorientation of homologs and prevent merotelic attachment of unified sister kinetochores. During meiosis in *C. elegans*, the site of the crossover, and not a localized centromere, ultimately determines the plane of chromosome orientation. During metaphase, the short arms of bivalents are lined up along the metaphase plate and the long arms point toward opposite poles. The kinetochore "cups" the entire volume of sister chromatids along the long arms. The activities that monoorient sisters and biorient homologs are concentrated at opposite domains of the bivalents. REC-8 promotes sister chromatid monoorientation along the long arms. By contrast, Aurora B is restricted to the short arm, where by preventing microtubules from "crossing the plane" of the bivalent short arm, it can ensure that sisters stay together, and homologs stay apart.

One way to conceptualize the differences between holocentric and monocentric chromosome orientation in meiosis is to think of holocentric bivalents as one large centromere without extended chromosome arms. The entire holocentric chromosome is cupped by the kinetochore. One domain within the kinetochore-bound domain (the long arm) contains REC-8, and according to this analogy, is comparable to core centromeric regions of monocentric chromosomes. Another domain (the short arm) is comparable to regions of Aurora B enrichment at centromeres. The organization of holocentric chromosomes magnifies the distinction between the various specialized chromosomal domains and makes it an ideal setting in which to examine these differences.

8. References

Andrews, P. D., Y. Ovechkina, et al. (2004). Aurora B regulates MCAK at the mitotic centromere. *Dev Cell* 6(2): 253-268.

Bernad, R., P. Sanchez, et al. (2011). Xenopus HJURP and condensin II are required for CENP-A assembly. *J Cell Biol* 192(4): 569-582.

Bernard, P., J. F. Maure, et al. (2001). Requirement of heterochromatin for cohesion at centromeres. *Science* 294(5551): 2539-2542.

Bhat, M. A., A. V. Philp, et al. (1996). Chromatid segregation at anaphase requires the barren product, a novel chromosome-associated protein that interacts with Topoisomerase II. *Cell* 87(6): 1103-1114.

Biggins, S. and A. W. Murray (2001). The budding yeast protein kinase Ipl1/Aurora allows the absence of tension to activate the spindle checkpoint. *Genes Dev* 15(23): 3118-3129.

Biggins, S., F. F. Severin, et al. (1999). The conserved protein kinase Ipl1 regulates microtubule binding to kinetochores in budding yeast. *Genes Dev* 13(5): 532-544.

Brito, I. L., H. G. Yu, et al. (2010). Condensins promote coorientation of sister chromatids during meiosis I in budding yeast. *Genetics* 185(1): 55-64.

Carvalho, A., M. Carmena, et al. (2003). Survivin is required for stable checkpoint activation in taxol-treated HeLa cells. *J Cell Sci* 116(Pt 14): 2987-2998.

Chan, R. C., A. F. Severson, et al. (2004). Condensin restructures chromosomes in preparation for meiotic divisions. *J Cell Biol* 167(4): 613-625.

Cheeseman, I. M., S. Anderson, et al. (2002). Phospho-regulation of kinetochore-microtubule attachments by the Aurora kinase Ipl1p. *Cell* 111(2): 163-172.

Cheeseman, I. M., J. S. Chappie, et al. (2006). The conserved KMN network constitutes the core microtubule-binding site of the kinetochore. *Cell* 127(5): 983-997.

Chelysheva, L., S. Diallo, et al. (2005). AtREC8 and AtSCC3 are essential to the monopolar orientation of the kinetochores during meiosis. *J Cell Sci* 118(Pt 20): 4621-4632.

Ciferri, C., S. Pasqualato, et al. (2008). Implications for kinetochore-microtubule attachment from the structure of an engineered Ndc80 complex. *Cell* 133(3): 427-439.

Cimini, D. (2007). Detection and correction of merotelic kinetochore orientation by Aurora B and its partners. *Cell Cycle* 6(13): 1558-1564.

Cimini, D., X. Wan, et al. (2006). Aurora kinase promotes turnover of kinetochore microtubules to reduce chromosome segregation errors. *Curr Biol* 16(17): 1711-1718.

Cobbe, N., E. Savvidou, et al. (2006). Diverse mitotic and interphase functions of condensins in Drosophila. *Genetics* 172(2): 991-1008.

Coelho, P. A., J. Queiroz-Machado, et al. (2003). Condensin-dependent localisation of topoisomerase II to an axial chromosomal structure is required for sister chromatid resolution during mitosis. *J Cell Sci* 116(Pt 23): 4763-4776.

Collette, K. S., Petty, E. L., et al., Different roles for Aurora B in condensin targeting during mitosis and meiosis. (2011). *J Cell Sci*: in press

Csankovszki, G., K. Collette, et al. (2009). Three distinct condensin complexes control C. elegans chromosome dynamics. *Curr Biol* 19(1): 9-19.

d'Erfurth, I., S. Jolivet, et al. (2009). Turning meiosis into mitosis. *PLoS Biol* 7(6): e1000124.

de Carvalho, C. E., S. Zaaijer, et al. (2008). LAB-1 antagonizes the Aurora B kinase in C. elegans. *Genes Dev* 22(20): 2869-2885.

Dej, K. J., C. Ahn, et al. (2004). Mutations in the Drosophila condensin subunit dCAP-G: defining the role of condensin for chromosome condensation in mitosis and gene expression in interphase. *Genetics* 168(2): 895-906.

DeLuca, J. G., W. E. Gall, et al. (2006). Kinetochore microtubule dynamics and attachment stability are regulated by Hec1. *Cell* 127(5): 969-982.

Dewar, H., K. Tanaka, et al. (2004). Tension between two kinetochores suffices for their bi-orientation on the mitotic spindle. *Nature* 428(6978): 93-97.

Ditchfield, C., V. L. Johnson, et al. (2003). Aurora B couples chromosome alignment with anaphase by targeting BubR1, Mad2, and Cenp-E to kinetochores. *J Cell Biol* 161(2): 267-280.

Dumont, J., K. Oegema, et al. (2010). A kinetochore-independent mechanism drives anaphase chromosome separation during acentrosomal meiosis. *Nat Cell Biol* 12(9): 894-901.

Gerlich, D., T. Hirota, et al. (2006). Condensin I stabilizes chromosomes mechanically through a dynamic interaction in live cells. *Curr Biol* 16(4): 333-344.

Goldstein, L. S. (1981). Kinetochore structure and its role in chromosome orientation during the first meiotic division in male D. melanogaster. *Cell* 25(3): 591-602.

Hagstrom, K. A., V. F. Holmes, et al. (2002). C. elegans condensin promotes mitotic chromosome architecture, centromere organization, and sister chromatid segregation during mitosis and meiosis. *Genes Dev* 16(6): 729-742.

Hartl, T. A., S. J. Sweeney, et al. (2008). Condensin II resolves chromosomal associations to enable anaphase I segregation in Drosophila male meiosis. *PLoS Genet* 4(10): e1000228.

Hauf, S., A. Biswas, et al. (2007). Aurora controls sister kinetochore mono-orientation and homolog bi-orientation in meiosis-I. *EMBO J* 26(21): 4475-4486.

Hauf, S., R. W. Cole, et al. (2003). The small molecule Hesperadin reveals a role for Aurora B in correcting kinetochore-microtubule attachment and in maintaining the spindle assembly checkpoint. *J Cell Biol* 161(2): 281-294.

Hauf, S. & Watanabe, Y. (2004). Kinetochore Orientation in Mitosis and Meiosis. *Cell* 119:317-327.

Hirano, T. (2000). Chromosome cohesion, condensation, and separation. *Annu Rev Biochem* 69: 115-144.

Hirano, T. (2002). The ABCs of SMC proteins: two-armed ATPases for chromosome condensation, cohesion, and repair. *Genes Dev* 16(4): 399-414.

Hirano, T. (2004). Chromosome shaping by two condensins. *Cell Cycle* 3(1): 26-28.

Hirano, T. and T. J. Mitchison (1994). A heterodimeric coiled-coil protein required for mitotic chromosome condensation in vitro. *Cell* 79(3): 449-458.

Hirota, T., D. Gerlich, et al. (2004). Distinct functions of condensin I and II in mitotic chromosome assembly. *J Cell Sci* 117(26): 6435-6445.

Honda, R., R. Korner, et al. (2003). Exploring the functional interactions between Aurora B, INCENP, and survivin in mitosis. *Mol Biol Cell* 14(8): 3325-3341.

Hudson, D. F., K. M. Marshall, et al. (2009). Condensin: Architect of mitotic chromosomes. *Chromosome Res* 17(2): 131-144.

Hudson, D. F., P. Vagnarelli, et al. (2003). Condensin is required for nonhistone protein assembly and structural integrity of vertebrate mitotic chromosomes. *Dev Cell* 5(2): 323-336.

Jessberger, R. (2002). The many functions of SMC proteins in chromosome dynamics. *Nat Rev Mol Cell Biol* 3(10): 767-778.

Kaitna, S., P. Pasierbek, et al. (2002). The aurora B kinase AIR-2 regulates kinetochores during mitosis and is required for separation of homologous Chromosomes during meiosis. *Curr Biol* 12(10): 798-812.

Kallio, M. J., M. L. McCleland, et al. (2002). Inhibition of aurora B kinase blocks chromosome segregation, overrides the spindle checkpoint, and perturbs microtubule dynamics in mitosis. *Curr Biol* 12(11): 900-905.

Kitajima, T. S., Kawashima, S. A. et al. (2004). The conserved kinetochore protein shugoshin protects centromeric cohesion during meiosis. *Nature* 427: 510-517.

Knowlton, A. L., W. Lan, et al. (2006). Aurora B is enriched at merotelic attachment sites, where it regulates MCAK. *Curr Biol* 16(17): 1705-1710.

Lampson, M. A., K. Renduchitala, et al. (2004). Correcting improper chromosome-spindle attachments during cell division. *Nat Cell Biol* 6(3): 232-237.

Lan, W., X. Zhang, et al. (2004). Aurora B phosphorylates centromeric MCAK and regulates its localization and microtubule depolymerization activity. *Curr Biol* 14(4): 273-286.

Lee, J., T. Miyano, et al. (2000). Specific regulation of CENP-E and kinetochores during meiosis I/meiosis II transition in pig oocytes. *Mol Reprod Dev* 56(1): 51-62.

Lee, J., S. Ogushi, et al. (2011). Condensin I and II are essential for construction of bivalent chromosomes in mouse oocytes. *Mol Biol Cell.*

Lens, S. M., R. M. Wolthuis, et al. (2003). Survivin is required for a sustained spindle checkpoint arrest in response to lack of tension. *EMBO J* 22(12): 2934-2947.

Lieb, J. D., M. R. Albrecht, et al. (1998). MIX-1: an essential component of the C. elegans mitotic machinery executes X chromosome dosage compensation. *Cell* 92(2): 265-277.

Liu, D., G. Vader, et al. (2009). Sensing chromosome bi-orientation by spatial separation of aurora B kinase from kinetochore substrates. *Science* 323(5919): 1350-1353.

Losada, A. and T. Hirano (2005). Dynamic molecular linkers of the genome: the first decade of SMC proteins. *Genes Dev* 19(11): 1269-1287.

Marston, A. L. and A. Amon (2004). Meiosis: cell-cycle controls shuffle and deal. *Nat Rev Mol Cell Biol* 5(12): 983-997.

Martinez-Perez, E., M. Schvarzstein, et al. (2008). Crossovers trigger a remodeling of meiotic chromosome axis composition that is linked to two-step loss of sister chromatid cohesion. *Genes Dev* 22(20): 2886-2901.

Monje-Casas, F., V. R. Prabhu, et al. (2007). Kinetochore orientation during meiosis is controlled by Aurora B and the monopolin complex. *Cell* 128(3): 477-490.

Moore, D. P. and T. L. Orr-Weaver (1998). Chromosome segregation during meiosis: building an unambivalent bivalent. *Curr Top Dev Biol* 37: 263-299.

Nabeshima, K., A. M. Villeneuve, et al. (2005). Crossing over is coupled to late meiotic prophase bivalent differentiation through asymmetric disassembly of the SC. *J Cell Biol* 168(5): 683-689.

Nasmyth, K. (2002). Segregating sister genomes: the molecular biology of chromosome separation. *Science* 297(5581): 559-565.

Nasmyth, K., J. M. Peters, et al. (2001). Splitting the chromosome: cutting the ties that bind sister chromatids. *Novartis Found Symp* 237: 113-133; discussion 133-118, 158-163.

Nonaka, N., T. Kitajima, et al. (2002). Recruitment of cohesin to heterochromatic regions by Swi6/HP1 in fission yeast. *Nat Cell Biol* 4(1): 89-93.

Oliveira, R. A., P. A. Coelho, et al. (2005). The condensin I subunit Barren/CAP-H is essential for the structural integrity of centromeric heterochromatin during mitosis. *Mol Cell Biol* 25(20): 8971-8984.

Ono, T., Y. Fang, et al. (2004). Spatial and temporal regulation of Condensins I and II in mitotic chromosome assembly in human cells. *Mol Biol Cell* 15(7): 3296-3308.

Ono, T., A. Losada, et al. (2003). Differential contributions of condensin I and condensin II to mitotic chromosome architecture in vertebrate cells. *Cell* 115(1): 109-121.

Page, S. L. & Hawley, R. S. (2003). Chromosome choreography: The meiotic ballet. *Science* 310: 785-789.

Paliulis, L. V. and R. B. Nicklas (2000). The reduction of chromosome number in meiosis is determined by properties built into the chromosomes. *J Cell Biol* 150(6): 1223-1232.

Parra, M. T., A. Viera, et al. (2004). Involvement of the cohesin Rad21 and SCP3 in monopolar attachment of sister kinetochores during mouse meiosis I. *J Cell Sci* 117(Pt 7): 1221-1234.

Parra, M. T., A. Viera, et al. (2003). Dynamic relocalization of the chromosomal passenger complex proteins inner centromere protein (INCENP) and aurora-B kinase during male mouse meiosis. *J Cell Sci* 116(Pt 6): 961-974.

Petersen, J., J. Paris, et al. (2001). The S. pombe aurora-related kinase Ark1 associates with mitotic structures in a stage dependent manner and is required for chromosome segregation. *J Cell Sci* 114(Pt 24): 4371-4384.

Petronczki, M., J. Matos, et al. (2006). Monopolar attachment of sister kinetochores at meiosis I requires casein kinase 1. *Cell* 126(6): 1049-1064.

Petronczki, M., M. F. Siomos, et al. (2003). Un menage a quatre: the molecular biology of chromosome segregation in meiosis. *Cell* 112(4): 423-440.

Rabitsch, K. P., M. Petronczki, et al. (2003). Kinetochore recruitment of two nucleolar proteins is required for homolog segregation in meiosis I. *Dev Cell* 4(4): 535-548.

Resnick, T. D., Satinover, D. L. et al., (2006) INCENP and Aurora B promote meiotic sister chromatid cohesion through localization of the Shugoshin MEI-S322 in Drosophila. *Dev Cell* 11: 57-68.

Ribeiro, S. A., J. C. Gatlin, et al. (2009). Condensin regulates the stiffness of vertebrate centromeres. *Mol Biol Cell* 20(9): 2371-2380.

Rogers, E., J. D. Bishop, et al. (2002). The aurora kinase AIR-2 functions in the release of chromosome cohesion in Caenorhabditis elegans meiosis. *J Cell Biol* 157(2): 219-229.

Ruchaud, S., M. Carmena, et al. (2007). Chromosomal passengers: conducting cell division. *Nat Rev Mol Cell Biol* 8(10): 798-812.

Saitoh, N., I. G. Goldberg, et al. (1994). ScII: an abundant chromosome scaffold protein is a member of a family of putative ATPases with an unusual predicted tertiary structure. *J Cell Biol* 127(2): 303-318.

Saka, Y., T. Sutani, et al. (1994). Fission yeast cut3 and cut14, members of a ubiquitous protein family, are required for chromosome condensation and segregation in mitosis. *EMBO J* 13(20): 4938-4952.

Sakuno, T., K. Tada, et al. (2009). Kinetochore geometry defined by cohesion within the centromere. *Nature* 458(7240): 852-858.

Sakuno, T. K. & Watanabe Y. (2009) Studies of meiosis disclose distinct roles of cohesion in the core centromere and pericentric regions. *Chromosome Res* 17:239-249.

Samoshkin, A., A. Arnaoutov, et al. (2009). Human condensin function is essential for centromeric chromatin assembly and proper sister kinetochore orientation. *PLoS ONE* 4(8): e6831.

Sandall, S., F. Severin, et al. (2006). A Bir1-Sli15 complex connects centromeres to microtubules and is required to sense kinetochore tension. *Cell* 127(6): 1179-1191.

Savvidou, E., N. Cobbe, et al. (2005). Drosophila CAP-D2 is required for condensin complex stability and resolution of sister chromatids. *J Cell Sci* 118(Pt 11): 2529-2543.

Schumacher, J. M., A. Golden, et al. (1998). AIR-2: An Aurora/Ipl1-related protein kinase associated with chromosomes and midbody microtubules is required for polar body extrusion and cytokinesis in Caenorhabditis elegans embryos. *J Cell Biol* 143(6): 1635-1646.

Schvarzstein, M., S. M. Wignall, et al. (2010). Coordinating cohesion, co-orientation, and congression during meiosis: lessons from holocentric chromosomes. *Genes Dev* 24(3): 219-228.

Severson, A. F., L. Ling, et al. (2009). The axial element protein HTP-3 promotes cohesin loading and meiotic axis assembly in C. elegans to implement the meiotic program of chromosome segregation. *Genes Dev* 23(15): 1763-1778.

Shang, C., T. R. Hazbun, et al. (2003). Kinetochore protein interactions and their regulation by the Aurora kinase Ipl1p. *Mol Biol Cell* 14(8): 3342-3355.

Siddiqui, N. U., P. E. Stronghill, et al. (2003). Mutations in Arabidopsis condensin genes disrupt embryogenesis, meristem organization and segregation of homologous chromosomes during meiosis. *Development* 130(14): 3283-3295.

Sonoda, E., T. Matsusaka, et al. (2001). Scc1/Rad21/Mcd1 is required for sister chromatid cohesion and kinetochore function in vertebrate cells. *Dev Cell* 1(6): 759-770.

Stear, J. H. and M. B. Roth (2002). Characterization of HCP-6, a C. elegans protein required to prevent chromosome twisting and merotelic attachment. *Genes Dev* 16(12): 1498-1508.

Steffensen, S., P. A. Coelho, et al. (2001). A role for Drosophila SMC4 in the resolution of sister chromatids in mitosis. *Curr Biol* 11(5): 295-307.

Strunnikov, A. V., E. Hogan, et al. (1995). SMC2, a Saccharomyces cerevisiae gene essential for chromosome segregation and condensation, defines a subgroup within the SMC family. *Genes Dev* 9(5): 587-599.

Tanaka, T., J. Fuchs, et al. (2000). Cohesin ensures bipolar attachment of microtubules to sister centromeres and resists their precocious separation. *Nat Cell Biol* 2(8): 492-499.

Tanaka, T. U. (2002). Bi-orienting chromosomes on the mitotic spindle. Curr Opin *Cell Biol* 14(3): 365-371.

Tanaka, T. U., N. Rachidi, et al. (2002). Evidence that the Ipl1-Sli15 (Aurora kinase-INCENP) complex promotes chromosome bi-orientation by altering kinetochore-spindle pole connections. *Cell* 108(3): 317-329.

Tomonaga, T., K. Nagao, et al. (2000). Characterization of fission yeast cohesin: essential anaphase proteolysis of Rad21 phosphorylated in the S phase. *Genes Dev* 14(21): 2757-2770.

Toth, A., K. P. Rabitsch, et al. (2000). Functional genomics identifies monopolin: a kinetochore protein required for segregation of homologs during meiosis i. *Cell* 103(7): 1155-1168.

Vagnarelli, P. and W. C. Earnshaw (2004). Chromosomal passengers: the four-dimensional regulation of mitotic events. *Chromosoma* 113(5): 211-222.

Vagnarelli, P., D. F. Hudson, et al. (2006). Condensin and Repo-Man-PP1 co-operate in the regulation of chromosome architecture during mitosis. *Nat Cell Biol* 8(10): 1133-1142.

Watanabe, Y. (2010). Temporal and spatial regulation of targeting aurora B to the inner centromere. *Cold Spring Harb Symp Quant Biol* 75: 419-423.

Watanabe, Y. and P. Nurse (1999). Cohesin Rec8 is required for reductional chromosome segregation at meiosis. *Nature* 400(6743): 461-464.

Watrin, E. and V. Legagneux (2005). Contribution of hCAP-D2, a non-SMC subunit of condensin I, to chromosome and chromosomal protein dynamics during mitosis. *Mol Cell Biol* 25(2): 740-750.

Wignall, S. M., R. Deehan, et al. (2003). The condensin complex is required for proper spindle assembly and chromosome segregation in Xenopus egg extracts. *J Cell Biol* 161(6): 1041-1051.

Wignall, S. M. and A. M. Villeneuve (2009). Lateral microtubule bundles promote chromosome alignment during acentrosomal oocyte meiosis. *Nat Cell Biol* 11(7): 839-844.

Winey, M., G. P. Morgan, et al. (2005). Three-dimensional ultrastructure of Saccharomyces cerevisiae meiotic spindles. *Mol Biol Cell* 16(3): 1178-1188.

Yokobayashi, S. and Y. Watanabe (2005). The kinetochore protein Moa1 enables cohesion-mediated monopolar attachment at meiosis I. *Cell* 123(5): 803-817.

Yokobayashi, S., M. Yamamoto, et al. (2003). Cohesins determine the attachment manner of kinetochores to spindle microtubules at meiosis I in fission yeast. *Mol Cell Biol* 23(11): 3965-3973.

Yu, H. G. and R. K. Dawe (2000). Functional redundancy in the maize meiotic kinetochore. *J Cell Biol* 151(1): 131-142.

Yu, H. G. and D. Koshland (2005). Chromosome morphogenesis: condensin-dependent cohesin removal during meiosis. *Cell* 123(3): 397-407.

Yu, H. G. and D. Koshland (2007). The Aurora kinase Ipl1 maintains the centromeric localization of PP2A to protect cohesin during meiosis. *J Cell Biol* 176(7): 911-918.

Yu, H. G. and D. E. Koshland (2003). Meiotic condensin is required for proper chromosome compaction, SC assembly, and resolution of recombination-dependent chromosome linkages. *J Cell Biol* 163(5): 937-947.

Zhu, C., E. Bossy-Wetzel, et al. (2005). Recruitment of MKLP1 to the spindle midzone/midbody by INCENP is essential for midbody formation and completion of cytokinesis in human cells. *Biochem J* 389(Pt 2): 373-381.

Epigenetics of the Synaptonemal Complex

Abrahan Hernández-Hernández[1],
Rosario Ortiz Hernádez[2] and Gerardo H. Vázquez-Nin[2]
[1]*Department of Cell and Molecular Biology, Karolinska Institute, Stockholm,*
[2]*Laboratory of Electron Microscopy, Faculty of Sciences,*
National Autonomous University of Mexico,
[1]*Sweden*
[2]*Mexico*

1. Introduction

Meiosis is a process composed of two divisions of the germ line cells without an intervenient S phase, thus there is no duplication of DNA between the first and the second meiotic divisions. The first meiotic division begins with the pre-leptotene that is the stage where the DNA replicates and chromosomes prepare to enter the meiotic prophase I in most of the organisms with sexual reproduction (Marston & Amon, 2004). The first meiotic division separates one homologue chromosome from the other member of the pair and in this way the two produced cells contain half number of chromosomes with two chromatids. During the metaphase of the first meiotic division the maternal and the paternal chromosomes of the bivalents are oriented at random, so the two new haploid cells receive a random number of chromosomes of each progenitor.

After a brief interphase the second division separates the sister chromatids of each chromosome and then the products of this division have the haploid number of chromosomes provided with one DNA double helix, frequently composed by segments of maternal and paternal DNA. In male mammals the final products of meiosis are four spermatids with half of the number of chromosomes of the species, with only one DNA double helix. However, in female organisms the final product of meiosis is one haploid oocyte and two small cells with a nucleus and a very small cytoplasm called polar bodies, which are not viable. During fecundation the union of two haploid gametes, the oocyte and the spermatozoa, recreates a diploid cell.

One of the biological significances of the meiosis is the production of genetic variability by the exchange of DNA between homologous chromosomes. Such exchange takes place during an extended meiotic prophase I and in most of the organisms proper meiotic genetic exchange depends on the accurate formation of a proteic structure between the homologous chromosomes, the synaptonemal complex (SC, for detailed revision see: Zickler & Kleckner 1998, 1999; Page & Hawley, 2004). There are several models used to study the SC and its importance for meiotic recombination, including yeast, Drosophila, C. elegans, plants and mice. Each model has its advantages and disadvantages. The mouse system has some advantages despite the low speed of genetics. Mouse genome and hence its chromosomes are larger than those of yeast and flies. This makes immunocytochemical analysis more powerful than in many other model organisms. Mouse genetics has been effectively used in combination with cytology to examine meiotic phenotypes produced as a result of targeted mutagenesis in embryonic stem cells.

2. Meiotic prophase I

Meiotic prophase I has been divided in five stages according to the chromosome morphology, meiotic recombination progression and the SC assembling. The interplay among these processes has been widely reviewed in different organisms (see Zickler & Kleckner 1998, 1999; Page & Hawley, 2004; Handel & Schimenti, 2010). Recently, with the identification of the histone code and its importance in gene regulation, chromatin structure and nuclear architecture (Turner, 2000; Jenuwein & Allis, 2001), old questions regarding chromosome structure and SC formation could be addressed. In this chapter we will focus our interest on the chromatin structure driven by epigenetic modifications and its relevance for SC formation and establishment, especially in mammals.

2.1 From chromosome homology recognition to synapsis

During the period G2 following meiotic phase S (some times called pre-leptotene) begins the recognition of similar sequences in the extended chromatin of homologous chromosomes. This process of recognition continues in leptotene and zygotene stages in microlampbrush chromosomes (Fig. 1).

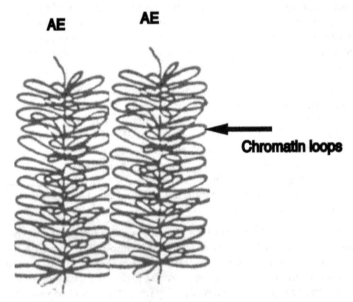

Lampbrush chromosomes

Fig. 1. Schematic drawing of the structure of homologous chromosomes in leptotene and zygotene stages of meiotic prophase during the process of alignment and recognition.

During pre-leptotene and leptotene the homologous chromosomes are not necessarily close to each other until the formation of the bouquet. The bouquet is a process that takes place during zygotene stage of the first meiotic prophase, the telomeres of the chromosomes slide associated to the nuclear membrane until they group in an area near the place of where the centrioles are located in the cytoplasm. In this way, the proximity of the chromosomes facilitates the recognition of homologies (Fig. 2)

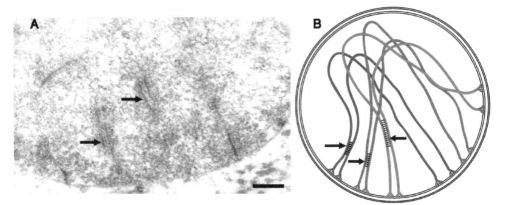

Fig. 2. A. - Electron microscopy image showing the pairing between homologous chromosomes (arrows) during the bouquet array and beginning of the formation of the synaptonemal complexes in zygotene stage. B. - Schematic representation of bouquet formation in zygotene stage. Homologous chromosomes migrate anchored to nuclear envelope until become close to each other, forming an array know as a bouquet. Synapsis takes place between homologous that are close enough to pair (arrows in A and B). Bar 500 nm.

The nature of the molecular mechanisms for this recognition of homologue sequences is not known, however, there are at least two processes proposed, one dependent on the distribution of transcription sites or factories (Cook, 1997) and the other dependent of non-spliced nascent RNA (Vázquez-Nin et al., 2003). According to the first view each chromosome has a unique array of transcription units along its length. Therefore, the chromatin fibrils with polymerases and transcription factors are folded into an array of loops, only the homologous chromosomes share similar distribution of loops with transcription factories and become zipped together (Cook, 1997). The second proposition also involves transcription as a possible mechanism of homology recognition. The study of the meiotic S phase (pre-leptotene), as well as leptotene and zygotene stages of meiotic prophase -that is the period of homology recognition and pairing- demonstrated an intense transcription but a very reduced pre-mRNA splicing. In this condition the newly synthesized mRNA could not be exported to the cytoplasm, as was demonstrated by quantitative autoradiography. So the function of newly synthesized mRNA must be inside the nucleus and in this period the main functions that were taking place inside the nuclei were homology recognition and pairing. Furthermore, electron microscope studies demonstrated a micro lampbrush structure of the chromosomes, which are in intense transcriptional activity. In pre-leptotene some loops of the micro lampbrush chromosomes contact loops of other chromosomes and the first parallel alignments of chromosomes take place (Fig. 1). Therefore, it was proposed that homologous chromosomes are in physical contact already at pre-leptotene stage (Vázquez-Nin et al., 2003). However it has not been shown a direct relationship of this chromosomal array with homology recognition.

During zygotene, as homologous chromosomes become aligned in pairs, the proteins of the lateral elements of the synaptonemal complex are incorporated to the chromosomal axis and the loops located between the axes leave the inter-axial space creating a region without DNA, which is the precursor of the central space of the synaptonemal complex.

3. The synaptonemal complex

The SC is a tripartite structure, which was described by Moses (1956) in spermatocytes of the crayfish. Since then it was found in all eukaryotic kingdoms (see reviews by Moses, 1956, 1968, 1969; Sotelo, 1969; Westergaard & von Wettstein 1972; Gillies, 1975; Loidl, 1990, 1991). SC morphology has been studied by means of electron microscopy. It is composed by two lateral elements (LEs) and a central region (CR). Each replicated homologous chromosome is anchored to one LE (Fig. 3), while completion of meiotic recombination (referred as crossover) takes place at late recombination nodules (RN) that are located in the CR (Fig. 3).

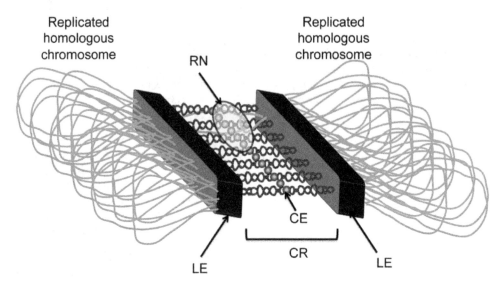

Fig. 3. Schematic representation of synaptonemal complex. The replicated homologous chromosomes are anchored to the lateral elements (LEs) of the synaptonemal complex (SC) while the genetic exchange between these homologous (referred as crossing over) takes place at the late recombination nodule (RN), that is tethered to the central region (CR).

The major protein components of the LEs are the meiosis specific proteins SYCP2 and SYCP3 (Dobson et al., 1994; Lammers et al., 1994; Offenberg et al., 1998; Winkel et al., 2009) as well as cohesion complexes (described below), whereas the CR is composed by SYCP1, SYCE1, SYCE2, SYCE3 and Tex12 (Figure 3) (Meuwissen et al., 1992; Costa et al., 2005; Hamer et al., 2008; Schramm et al., 2011). SYCP3 is a major structural component of vertebrate synaptonemal complexes. The evolutionary conserved domains of SYCP3, the alpha helix together with two flanking motif CM1 and CM2, are necessary and sufficient for SYCP3 polymerization and assembly of high order structures (Baier et al., 2007). Nevertheless, some differences in the SYCP3 expression have been found among mammals. In contrast to other vertebrates, rat and mouse SYCP3 exists in two isoforms. The short isoform is conserved among vertebrates. However, the longer isoform, which represents an N-terminal extension of the shorter one, most likely appeared about 15 millions years ago in a common ancestor of rat and mouse and after the separation of the hamster branch (Alsheimer et al., 2010). SYCP2 and SYCP3 incorporate to the axial elements of chromosomes during lepto-zygotene, forming fibrous cores in the homologous chromosomes (Hamer et al., 2008 and references therein).

The C-terminus of SYCP1 directly interacts with SYCP2 (Winkel et al., 2009). These authors proposed that SYCP2 acts as a linker between SYCP1 and SYCP3 and therefore it could be the connecting link between lateral elements and transverse filaments of the CR (Winkel et al., 2009). On the other hand, the N-terminus of SYCP1 is associated in the middle of the CR with another N-terminus of SYCP1 forming the central element (CE) of the CR. At the CE are also found the proteins SYCE1, SYCE2, SYCE3 and Tex12, which are important for the proper CR assembling and for the crossover resolution (Bolcun-Filas et al., 2007, 2009; Hamer et al., 2008, Schramm et al., 2011). Defects in the organization of the synaptonemal complex result in alterations of the meiotic recombination and infertility.

Cohesins are chromosomal proteins that form complexes involved in the maintenance of sister chromatid cohesion during division of somatic and germ cells. Three meiotic cohesins subunits have been reported in mammals, REC8, STAG3 and SMC1 beta, their expression has been found in mouse spermatocytes (Prieto et al., 2004) and human oocytes (García-Cruz et al., 2010). SMC1 beta, SMC3 and STAG3 are localized along axial fibers of leptotene-zygotene chromosomes and then to the LE of the SC. Cohesins are essential for completion of recombination, pairing, meiotic chromosome axis formation, and assembly of the SC. Rec8 is involved in several functions as cohesion, pairing, recombination, chromosome axis and SC assembly (Brar et al., 2009). At difference from meiosis in male mice, the cohesin axis is progressively lost in oocytes, with parallel destruction of the axial elements at dictyated arrest (Prieto et al., 2004).

The SC is important for the normal formation of crossovers. In many, but not all, organisms, the homology search, that occurs mediated by DNA-DNA interactions, is also intimately associated with the movement of homologous chromosomes to bring them into close juxtaposition (Székvölgyi & Nicolas, 2010).

Chiasmata formed by crossovers are central for the process of chromosome segregation as they hold together the homologous chromosomes at metaphase of the first meiotic division; at least one crossover per pair of homologs allows that each member of the pair migrates to an opposite pole of the spindle (Székvölgyi & Nicolas, 2010).

Most eukaryotes possess two recombinases, Rad51 and Dmc1. Homologues of these proteins are widely conserved in nature, from virus to humans. In eukaryotes Rad51 is required for most homologous recombination pathways in both mitotic and meiotic cells (Kagawa & Kurumizaka, 2009 and references therein).

4. Chromosome organization on the synaptonemal complex

As mentioned above, the SC is essential for meiotic recombination completion; this is because crossover formation depends in the accurate formation of the CR of the SC. The crossovers are observed as dense structures associated to the CR, as showed by electron microscopy. Crossovers are referred as late recombination nodules (RN, Fig. 3) and they are observed tether to the CR of the SC (Carpenter, 1975, 1979, 1981). Therefore, the SC is considered as the scaffold to which the chromosomes are anchored while they exchange genetic material.

After DNA replication at pre-leptotene stage, replicated sister chromatids are held together at specific points by the cohesion component SMC3 and the meiosis-specific cohesin Rec8 is incorporated to this scaffold. As the cells progress to leptotene stage, the chromosomes undergo condensation and SMC1 beta/STAG3, other meiosis-specific cohesins, are incorporated to the cohesion scaffold. At this stage fine filaments formed by the cohesin scaffold can be identified by immunocytochemical approaches. These filaments, called axial elements (AEs) are the precursors of the LEs and are surrounded by chromatin loops protruding out of them (Fig. 1). In this stage, SYCP2 and SYCP3 are incorporated to the AEs and in zygotene stage SYCP1 and accessory proteins begin to synapse the aligned AEs. During pachytene the SC is fully formed throughout the whole length of LEs. The homologous chromosomes are anchored to the AEs in early stages and to the LEs in pachytene stage. However the mechanism of association of the chromosomes to the AEs/LEs has been controversial and poorly understood.

The presence of DNA in the LEs was documented by enzymatic digestion followed by staining methods almost at the same time as the SC was observed (Coleman & Moses, 1964) and corroborated by immunocytochemical approaches in later studies (Vázquez-Nin et al., 1993). There have been few studies trying to identify DNA sequences associated to the LEs of the SC. In C. elegans, the chromosomes pair at specific areas known as pairing centers, recently it has been shown that repeat sequences motifs are at these pairing centers (Phillips et al., 2009). It has been suggested that in mammals repeat sequences interspersed through the genome, are responsible to anchor the chromosomes to the LEs (Pearlman et al., 1992; Hernández-Hernández et al., 2008). However not all the bulk of repeat sequences are incorporated into the LEs, suggesting a mechanism of selection of specific sequences to be anchored to the LE. Further experiments have shown that the chromatin structure at these lateral element-associated repeated sequences (LEARS) is in part responsible for their association to the SC (Hernández-Hernández et al., 2010).

4.1 Chromosomes anchor to the lateral elements by means of specific DNA sequences

The presence of DNA in the inner part of the LEs suggested that chromosomes are anchored by means of specific sequences. Two different studies have shown that LEs contain specific DNA sequences. One of the studies suggested that LEs associate DNA consist in repeat sequences like long and short interspersed elements (LINE/SINE) (Pearlman et al., 1992). In the second study the authors used chromatin immunoprecipitation (ChIP) using anti-sycp3 antibody to pull down DNA sequences associated to the LE (Hernández-Hernández et al., 2008). All the immunoprecipitated sequences consisted of repeat DNA, like LINE, SINE, long terminal repeats (LTR), satellite, and simple repeats. The presence of these sequences in the LEs has been corroborated by means of DNA in situ hybridization at the optical and electron microscope level (Hernández-Hernández et al., 2008; Spangenberg et al., 2010). Therefore, these specific sequences have been called lateral element-associated repeat sequences or LEARS. However the presence of some other sequences can not be ruled out with these studies, more analysis are needed to determinate whether these are the only sequences helping the chromosomes to anchor to the LEs.

5. Chromatin structure in the LEs of the SC

Chromatin immunoprecipitation experiment using the LEs specific protein SYCP3, demonstrated enrichment of repeat DNA sequences, which localize to the LEs, as well as in the bulk of the chromatin, as shown by in situ hybridization (Hernández-Hernández et al., 2008). However, features in the primary structure of LEARS did not reveal any obvious consensus sequence, suggesting that secondary structure might be responsible for recruitment of LEARS to the LEs. In somatic cells, most of these transcriptionally inactive repeat sequences are subject to epigenetic modifications favoring their organization in heterochromatin (Martens et al., 2005). Furthermore, chromatin structure dictated by epigenetic modifications during meiosis is critical for accurate SC assembly and meiosis progression (Hernández-Hernández et al., 2009). Therefore, it is possible that specific epigenetic modifications of LEARS influence their interaction with LEs. To address this hypothesis our group has performed immunofluorescent detection (IF) of histone post translation modifications (PTM) that are associated to repeat sequences in somatic cells (Martens et al., 2005). We found specific association of PTM with the SC during pachytene stage. Tri-methylation of histone H3 on lysine 9 and tri-methylation of histone H4 on lysine 20 (H3K9me3 and H4K20me3 respectively) co-localize with one extreme of the SCs (Fig. 4), whereas tri-methylation of histone H3 on lysine 27 (H3K27me3) co-localizes to the SC in almost all its length (Fig. 4). We then followed the dynamics of co localization of these specific marks throughout the meiotic prophase I.

Leptotene stage: at this stage, sycp3 antibody stains fine filaments that correspond to the axial elements. H3K9me3 and H4K20me3 are already co-localizing with one of the extremities of the AE. Centromeres are located close to the end of acrocentric chromosomes and in rat they are mainly composed of minor and mayor satellite DNA repeats. These sequences are enriched with the PTMs H3K9me3 and H4K20me3 (Martens et al., 2005). In cells undergoing meiosis and SC assembling, the centromeres are located near one of the extremes of the AE nearby the nuclear envelope. The staining pattern of H3K9me3 and H4K20me3 in the AE therefore may correspond to the satellite repeats present in the centromeric and pericentromeric region. H3K27me3 was absent from the whole nucleus at this stage.

Zigotene: the AEs of homologous chromosomes start to synapse. H3K9me3 and H4K20me3 are located in the extreme of the SC in formation.

Pachytene: The SC between homologous chromosomes is completely formed. H3K9me3 and H4K20me3 continue associated to one of the extremes of the SC (Fig. 4). In this stage

H3K27me3 staining pattern is visible and this PTM is co-localizing with SYCP3 throughout patches of the SC (Fig. 4). The results of these IFs suggest that these three histone marks may be involved in the chromatin structure at the LEARS in the LEs.

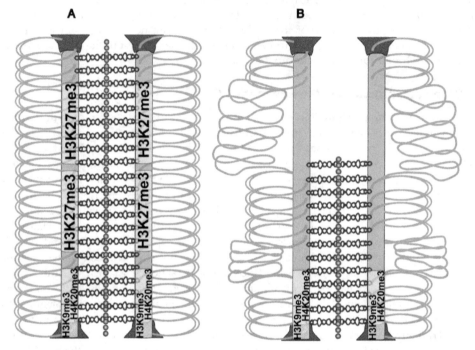

Fig. 4. Model of chromatin structure in the LEs of the SC. A. – Homologous chromosomes are attached to the LEs of the SC by means of specific DNA sequences (LEARS). These anchorage DNA sequences are enriched with histone PTMs that determinate their chromatin structure. B. – Blocking of chromatin structure leads to a defective chromosome anchorage to the LEs. H3K27me3 is not longer associated with SINE and LTR sequences producing their detachment from the LEs.

5.1 Epigenetic profile of LEARS

In order to understand whether the specific PTM are associated with the LEARS, we performed two rounds of ChIP assays (re-ChIP). The first round of ChIP analyses was done using the SYCP3 antibody to pull down the LEARS and the chromatin proteins associated to them. As a template for the second round of ChIP, we used the pulled down complexes from the first round of ChIP (SYCP3 and chromatin of LEARS) and pulled down DNA sequences associated with the distinct PTM of interest (H3K9me3, H3K27me3 and H4K20me3). Enrichment of the different LEARS with PTM marks was assessed by semi quantitative polymerase chain reaction (sqPCR) (Hernandez-Hernandez et al., 2010). Table 1 summarizes the enrichment of PTM in the different LEARS. Satellite repeats are enriched with H3K9me3 and H4K20me3, whereas LINE sequences are enriched with H4K20me3. SINE and LTR sequences are enriched with H3K27me3.

LEARS	Histone mark (PTM)		
	H3K9me3	H4K20me3	H3K27me3
LINE	-	Enriched	
SINE	-		Enriched
LTR	-		Enriched
Satellite	Enriched	Enriched	

Table 1. LEARS and their enrichment with histone marks

By analyzing the patterns of IF for PTMs/SC proteins during the meiotic prophase and the enrichment of PTM in the LEARS, we can predict whether or not histone PTMs have a role in the recruitment of LEARS to the LE of the SC.

H3K9me3 and H4K20me3 are already enriched in the satellite repeats of the centromeric region when the AE is formed during leptotene stage. Probably these marks are constitutive for satellite repeats, since in somatic cells this enrichment has been reported before (Marten et al., 2005). Therefore the likelihood that these PTMs are important for LEARS recruitment is scarce. LINE sequences are enriched with H4K20me3, the same PTM present in satellite sequences. However the staining pattern only resembles that for satellite sequences. LINE sequences are not confined to centromeric regions, rather they are present all along the chromosomes, therefore it would be expected that the staining pattern would be co localizing with SYCP3 not only at one of the extremes, but to the entire length of the SC. A possible explanation for the absence of IF signal for H3K20me3 in the whole length of the SC is that LINE sequences are not highly clustered as the satellite sequences in specific regions of the chromosome. The clustered satellite regions in one of the extremities of the SC produce a strong IF signal, making difficult to detect the signal of LINE sequences in regions were there is not clustering, for example along the LEs. Thus, LINE sequences are enriched with H4K20me3, but they are masked by the signal from satellite DNA regions.

5.2 A specific histone mark appears at the time the SC is mature

H3K27me3 colocalizes with SYCP3 along patches through the whole length of the SC. Strikingly; this pattern is only seen in pachytene stage when the SC is mature (Hernández-Hernández et al., 2010). Furthermore, this histone mark was enriched in SINE- and LTR-LEARS. These evidences suggest that SINE and LTR sequences are enriched with H3K27me3, conferring a specific chromatin structure that in turn is contributing to the LE structure, probably anchoring and/or maintaining the attachments of these sequences to the LEs.

To address this hypothesis, we decided to treat rat testicles with tricostatin-A (TSA), a histone desacetylase (HDACs) inhibitor, during nine days. After this period of treatment, pachytene cells have received the treatment since leptotene stage, according to the duration of meiotic prophase I (Adler, 1996). TSA has been shown to effectively inhibit HDACs, blocking downstream reactions and finally methylation of histone lysine residues (Ekwall et al., 1997). Then we perform IF, ChIP and re-ChIP experiments to assess the effect of HDACs inhibition on the SC structure (Hernández-Hernández et al., 2010).

H3K9me3 and H4K20me3 IF signal was reduced at pachytene stage, but the pattern is the same as that of non-treated animals. Suggesting that most of these marks were deposited on centromers before leptotene stage. In agreement with this, leptotene and zygotene cells, which have been treated since they were at pre-leptotene stages, showed reduced signal and more scattered IF pattern that in non-treated animals. TSA treatment therefore, partially affected deposition of these PTMs and chromatin structure as well. Enrichment of H3K9me3 and H4K20me3 in satellite sequences and H4K20me3 in LINE sequences was significantly reduced. However these two classes of LEARS are still attached to the LEs of the SC as shown by ChIP assays in treated and control animals (Hernández-Hernández et al., 2010). In summary, TSA affected enrichment of H3K9me3 and H4K20me3 in satellite and LINE sequences, but this loss of enrichment does not produce loss of these two LEARS from the LE of the SC. Implying that satellite and LINE sequences may have another unidentified PTM that is involved in their association to the LE or that these sequences do not associate to the LE via histone modifications. Moreover, staining pattern in treated animal at one of the extremes of the EA/LE, suggest that most of this mark was not altered during such a short period of treatment (Hernández-Hernández et al., 2010).

The most striking result was the observed for H3K27me3. The staining pattern for H3K27me3 is specific for pachytene cells, co localizing with SYCP3 along stretches of the SC. This PTM is also enriched in the SINE and LTR sequences attached to the LEs. After TSA treatment, the IF signal for H3K27me3 almost disappears from the nucleus of pachytene cells. Furthermore, the enrichment of this PTM in SINE and LTR was significantly reduced and these sequences were not longer associated to the LEs of the SC. These results suggest that H3K27me3 appears in pachytene stage and is important for attachment and/or maintenance of SINE and LTR sequences in the LEs o the SC (Hernández-Hernández et al., 2010).

5.3 Failures in pairing between homologous chromosomes and loss of DNA-associated to the LE

To evaluate the direct effect of loss of LEARS from the LEs of the SC, we analyzed the ultrastructure of the SC by means of optical and electron microscopy. IF staining of LEs and CR of the SC in sections allowed us to identify that homologous chromosomes are paired but synapsis is not complete. By means of electron microscopy we found a high incidence of LEs that were not synapsed. These defects in SC formation in turn activate the programmed cell death of spermatocytes in late pachytene stage observed in sections of seminiferous tubules. Furthermore, when homologous were partially synapsed, we identified failures in the CR structure. The CR was formed between the homologous but not throughout the entire length of the SC. By means of a specific ultrastructural DNA staining, we observed that the DNA pattern was less dense in the LEs of treated animals than in the LEs from non-treated rats (Hernández-Hernández et al., 2010). This suggested that DNA association to the LEs was altered and that the CR in not completely formed between the homologous chromosomes.

6. A model of the chromatin structure in the LEs of the SC

Taking in account all the evidences, we proposed that chromatin structure of a sub set of LEARS (SINE and LTR) is important for LEARS recruitment and/or maintenance in the LEs of the SC. Blocking of this specific chromatin structure leads to failures in SC structure, detachment of DNA sequences form the LEs (SINE, LTR DNA sequences) and finally to cell death. Therefore we suggest that H3K27me3 is an indispensable histone PTM important for chromosome attachment to the LEs and hence SC structure.

7. References

Adler, ID. (1996). Comparison of the duration of spermatogenesis between male rodents and humans. *Mutat Res,* 10,352(1-2):169-72.

Alsheimer, M., Baier, A., Schramm S., Shütz, W., & Benavente, R. (2010). Synaptonemal complex protein SYCP3 exists in two isoforms showing different conservation in mammalian evolution. *Cytogenetics and Genome Research,* 128(1-3):162-8.

Baier, A., Alsheimer, M., Voiff, JN., & Benavente, R. (2007). Synaptonemal complex protein SYCP3 of the rat: evolutionarily conserved domains and the assembly of higher order structures. *Sex Dev,* 1(3):161-168.

Bolcun-Filas, E., Costa, Y., Speed, R., Taggart, M., Benavente, R., De Rooij, DG., & Cooke, HJ. (2007). SYCE2 is required for synaptonemal complex assembly, double strand break repair, and homologous recombination. *J Cell Biol,* 176(6):741-7.

Bolcun-Filas, E., Hall, E., Speed, R., Taggart, M., Grey, C., de Massy, B., Benavente, R., & Cooke, HJ. (2009). Mutation of the mouse Syce1 gene disrupts synapsis and suggests a link between synaptonemal complex structural components and DNA repair. *PLoS Genet,* 5(2):e1000393.

Brar, GA., Hochwagen, A., Ee, LS., & Amon, A. (2009). The multiple roles of cohesin in meiotic chromosome morphogenesis and pairing. *Mol Biol Cell,* 20(3):1030-47.

Carpenter, ATC. (1975). Electron microscopy of meiosis in Drosophila melanogaster females: II: The recombination nodule-a recombination-associated structure at pachytene? *Proc Nat Acad Sci USA,* 72(8):3186-3189.

Carpenter, ATC. (1979). Synaptonemal complex and recombination nodules in wild-type Drosophila melanogaster females. *Genetics,* 92(2):511-541.

Carpenter, ATC. (1981). EM autoradiographic evidence that DNA synthesis occurs at recombination nodules during meiosis in Drosophila melanogaster females. *Chromosoma,* 83:59-80.

Coleman, JR., & Moses, MJ. (1964). DNA and the fine structure of synaptic chromosomes in the domestic rooster (gallus domesticus). *J Cell Biol,* 23:63-78.

Costa, Y., Speed, R., Ollinger, R., Alsheimer, M., Semple, CA., Gautier, P., Maratou, K., Novak, I., Höög, C., Benavente, R., & Cooke, HJ. (2005). Two novel proteins recruited by synaptonemal complex protein 1 (SYCP1) are at the center of meiosis. *J Cell Sci,* 118:2755-2762.

Cook, PR. (1997). The transcriptional basis of chromosome pairing. *J Cell Sci,* 110(9):1033-1040.

Dobson, MJ., Pearlman, RE., Karaiskakis, A., Spyropoulos, B., & Moens, PB. (2007). Synaptonemal complex proteins: occurrence, epitope mapping and chromosome disjunction. *J Cell Sci*, 107:2749-2760.

Ekwall, K., Olsson, T., Turner, BM., Cranston, G., & Allshire, RC. (1997). Transient inhibition of histone deacetylation alters the structural and functional imprint at fission yeast centromeres. *Cell*, 91:1021–1032

García-Cruz, R., Brieño, MA., Roig, I., Velilla, E., Pujol, A., Cabero, L., Pessarronda, A., Barbero, JL., & García Caldés, M. (2010). Dynamics of cohesin proteins REC8, STAG3, SMC1 beta and SMC3 are consistent with a role in sister chromatid cohesion during meiosis in human oocytes. *Hum Reprod*, 25:2316-2327.

Gillies, CB. (1975). Synaptonemal complex and chromosome structure. *Annu Rev Genet* 9:91-109.

Hamer, G., Wang, H., Bolcun-Filas, E., Cooke, HJ., Benavente, R., & Höög, C. (2008). Progression of meiotic recombination requires structural maturation of the central element of the synaptonemal complex. *J Cell Sci*, 121:2445-2451.

Handel MA., & Schimenti, JC. (2010) Genetics of mammalian meiosis: regulation, dynamics and impact on fertility. *Nat Rev Genet*, 11(2):124-36.

Hernández-Hernández, A., Rincón-Arano, H., Recillas-Targa, F., Ortiz, R., Valdes-Quezada, C., Echeverría, OM., Benavente, R., & Vázquez-Nin, GH. (2008). Differential distribution and association of repeat DNA sequences in the lateral element of the synaptonemal complex in rat spermatocytes. *Chromosoma*, 117(1):77-87.

Hernández-Hernández, A., Vázquez-Nin, GH., Echeverría, OM., & Recillas-Targa, F. (2009). Chromatin structure contribution to the synaptonemal complex formation. *Cell Mol Life Sci*, 66:1198-1208.

Hernández-Hernández, A., Ortiz, R., Ubaldo, E., Martínez, OM., Vázquez-Nin, GH., & Recillas-Targa, F. (2010). Synaptonemal complex stability depends on repressive histone marks of the lateral element-associated repeat sequences. *Chromosoma*, 119(1):41-58.

Jenuwein, T., & Allis, CD. (2001). Translating the histone code. *Science*, 293(5532):1074-80.

Kagawa, W., & Kurumizaka, H. (2009). From meiosis to postmeiotic events: Uncovering the molecular roles of the meiosis-specific recombinase Dmc1. *FEBS J*, 277(3):590-598.

Lammers, JH., Offenberg, HH., van Aalderen, M., Vink, AC., Dietrich, AJ., & Heyting, C. (1994). The gene encoding a major component of the lateral elements of synaptonemal complexes of the rat is related to X-linked lymphocyte-regulated genes. *Mol Cell Biol*, 14(2): 1137-1146.

Loidl, L. (1990). The initiation of meiotic pairing: the cytological view. *Genome*, 33: 759-778.

Loidl, L. (1991). Coming to grips with a complex matter. A multidisciplinary approach to the synaptonemal complex. *Chromosoma*, 100:289-292.

Marston, AL., & Amon, A. (2004). Meiosis: cell-cycle controls shuffle and deal. *Nat Rev Mol Cell Biol*, 5(12):983-997.

Martens, JH., O'Sullivan, RJ., Braunschweig, U., Opravil, S., Radolf, M., Steinlein, P., & Jenuwein, T. (2005). The profile of repeat-associated histone lysine methylation states in the mouse epigenome. *EMBO J*, 24(4):800-812.

Meuwissen, RL., Offenberg, HH., Dietrich, AJ., Riesewijk, A., van Iersel, M., & Heyting, C. (1992). A coiled-coil related protein specific for synapsed regions of meiotic prophase chromosomes. *EMBO J*, 11(13): 5091-5100.

Moses, MJ. (1956). Chromosomal structures in crayfish spermatocytes. *J Biophys Biochem Cytol*, 2:215-218.

Moses, MJ. (1968). Synaptonemal complex. *Annu Rev Genet*, 2:363-412.

Moses, MJ. (1969). Structure and function of the synaptonemal complex. *Genetics*, 67:41-51.

Offenberg, HH., Schalk, JA., Meuwissen, RL., van Aalderen, M., Kester, HA., Dietrich, AJ., & Heyting, C. (1998). SCP2: a major protein component of the axial elements of synaptonemal complexes of the rat. *Nuclei Acid Res*, 26:2572-2579.

Page, SL., & Hawley, RS. (2004). The genetics and molecular biology of the synaptonemal complex. *Annu Rev Cell Dev Biol*, 20:525-58.

Pearlman, RE., Tsao, N., & Moens, PB. (1992). Synaptonemal complexes from DNase-treated rat pachytene chromosomes contain (GT)n and LINE/SINE sequences. *Genetics*, 130(4):865-872.

Phillips, CM., Meng, X., Zhang, L., Chretien, JH., Urnov, FD., & Dernburg, AF. (2009). Identification of chromosome sequence motifs that mediate meiotic pairing and synapsis in C. elegans. *Nat Cell Biol*, 11(8):934-942.

Prieto, I., Tease, C., Pezzi, N., Buesa, JM., Ortega, S., Kremer, L., Martínez, A., Martínez-A, C., Hultén, MA., & Barbero, JL. (2004). Cohesin component dynamic during meiotic prophase I in mammalian oocytes. *Chromosome Res*, 12(3):197-213.

Schramm, S., Fraune, J., Naumann, R., Hernandez-Hernandez, A., Höög, C., Cooke, HJ., Alsheimer, M., & Benavente, R. (2011). A novel mouse synaptonemal complex protein is essential for loading of central element proteins, recombination, and fertility. *PLoS Genet*, 7(5):e1002088.

Sotelo, JR. (1969). Ultrastructure of the chromosomes in meiosis. In: *Handbook of Molecular Cytology*, Ed. Lima de Faria A, pp 412-434; 1969, North Holland Pub. Co. Amsterdam.

Székvölgyi, L. & Nicolas, A. (2010). From meiosis to postmeiotic events: Homologous recombination is obligatory but flexible. *FEBS J*, 277:571-589.

Spangenberg, VE., Dadashev, SIa., Matveevskii, SN., Kolomiets, OL., & Bogdanov, IuF. (2010). How do chromosomes attach to synaptonemal complexes? *Genetika*, 46(10):1363-6.

Turner, BM. (2000). Histone acetylation and an epigenetic code. *Bioessays*, 22(9):836-845.

Vázquez-Nin, GH., Flores, E., Echeverría, OM., Merkert, H., Wettstein, R., & Benavente, R. (1993). Immunocytochemical localization of DNA in synaptonemal complexes of rat and mouse spermatocytes, and chick oocytes. Chromosoma, 102(7):457-463.

Vázquez-Nin, GH., Echeverría OM., Ortiz, R., Scassellati, C., Martín, TE., Ubaldo, E., & Fakan, S. (2003). Fine structural cytochemical analysis of homologous chromosome recognition, alignment, and pairing in guinea pig spermatogonia and spermatocytes. *Biol Reprod*, 69(4):1362-1379.

Westergaard, M., & von Wettstein, D. (1972). The synaptonemal complex. Annu Rev Genet, 6:71-110.

Winkel, K., Alsheimer, M., & Benavente, R. (2009). Protein SYCP2 provides a link between transverse filaments and lateral elements of mammalian complexes. Chromosoma, 118(2): 259-267.

Zickler, D., & Kleckner, N. (1998). The leptotene-zygotene transition of meiosis. *Annu Rev Genet*, 32:619-97.

Zickler, D., & Kleckner, N. (1999). Meiotic chromosomes: integrating structure and function. *Annu Rev Genet*, 33:603-754.

3

Meiotic Irregularities in Interspecific Crosses Within Edible Alliums

Agnieszka Kiełkowska

*University of Agriculture in Krakow,
Poland*

1. Introduction

The economically most important edible alliums are onion (*Allium cepa* L.), Japanese bunching onion (*A. fistulosum* L.), leek (*A. apmleroprasum* spp. *porrum* L.), and garlic (*A. sativum* L.). These species are mostly used as condiments, but also have medicinal value (Keusgen, 2002). Onion and garlic are grown worldwide, leek is predominantly grown in Europe, and Japanese bunching onion in East Asia (Kik, 2002). Hybridization of certain genotypes, increasing genetic variability, is the predominant breeding method used for onion, leek and Japanese bunching onion. Garlic breeding is accelerated by clonal selection (Etoh & Simon, 2002).

Interspecific hybridization within the genus *Allium* has a long history, and has been an important tool for increasing genetic variation. It was used for creating new varieties as well as for transferring agronomically useful traits from wild relatives (Brewster, 1994; Kiełkowska & Adamus, 2010; Kik, 2002). However, techniques of sexual hybridization very often have been accompanied by several difficult to overcome pre- and post-fertilization barriers, resulting in a limited number of obtained F1 hybrids, as well as difficulties with back-crossing. Meiotic studies of hybrids and back-cross progenies within *Allium* revealed interesting information about the nature of alien introgression and improved our understanding of "foreign" chromatin transmission into cultivated species (Kik, 2002; van Heusden et al., 2000).

The purpose of the present paper is to give a concise review of meiosis in pollen mother cells (PMC) and pollen fertility in *Allium cepa* and its crosses with *A. fistulosum*, *A. ampeloprasum*, and *A. sativum*.

2. Chromosome number and meiosis

Most alliums are diploids (2n=2x), with basic chromosome numbers x=7 (North America), x=8 (Eurasia and Mediterranean basin), or x=9 (Eurasia) (Havey, 2002; Ved Brat, 1965a). Polyploids such as triploids (*A. rupestre*, *A. scordoprasum*), tetraploids (*A. ampeloprasum*, *A. chinense*, *A. nutans*), pentaploids (*A. splendens*), hexaploids (*A. lineare*), and octoploids (*A. nutans*) also occur (Jones R.N., 1990; Jones R.N & Rees, 1968; Ved Brat, 1965a).

Accessory chromosomes (B-chromosomes, supernumerary) have been documented (Bougourd and Plowman 1996). Their size and centromere position is variable within species (Ved Brat, 1965a, 1965b). B-chromosomes found in *A. paniculatum* never undergo pairing with other chromosomes during meiosis, however, occurrence of chiasmata associations among B-chromosomes in *A. schoenoprasum* was observed (Bougourd & Parker 1979).

The majority of species within the genus carry meta- or submetacentric chromosomes (Ved Brat, 1965a). Levan (1932, 1935) noted that species with x=7 had larger chromosomes than those with x=8 or x=9 and that arm-length asymmetry was more frequent in the "16" and "18"-chromosomes types and that the "14"-chromosome types were the most primitive. Ved Brat (1965a) estimated, that 40% of the forms within the genus possessing eight chromosomes had varying number of asymmetrical chromosomes, often carrying the nucleolar organizer regions (NOR), whereas remaining 60% had symmetrical chromosomes. Reports about differences in telomeric sequences in *Allium* (Fuchs et al., 1995) and stabilization of chromosome ends by highly repetitive satellites and rDNA (Pich et al., 1996) were also published.

Length of the alliums chromosomes also varies; the shortest (~7 µm) chromosomes were found in *A. yunnanense* (x=8), the longest (~22 µm) in *A. fragrans* (x=9) (Levan 1935). Investigation of the nuclear DNA content revealed that evolution of alliums was associated with variation in DNA amounts. Nuclear DNA content was proportional to chromosome 'volume' in the cell, and changes in the DNA amount were distributed between chromosomes within complements (Jones R.N. & Rees, 1968). *Alliums* with n=8 have in general the lowest; mean 28,9 pg/2C nuclear DNA amounts, while those with n=9, the highest; mean 42,2 pg/2C (Bennett & Leitch, 2010).

Breeding systems in alliums are variable; some are reproduced sexually as outbreeders or inbreeders, some asexually with vivapory or apomixis and some are vegetatively propagated (Ved Brat, 1965b). An overwhelming majority of the genus is propagated sexually, which allows for gene exchange and, in consequence, more genetic variability. However, in the vegetatively reproduced *A. sativum* the main source of variation is the occurrence of alterations in the number and morphology of chromosomes in somatic cells of the clone (Konvička & Levan, 1972).

Meiosis in alliums was reviewed by Levan (1931, 1935). In prophase I of the diploid forms chiasmata are visible at early diakinesis. Terminalization of chiasmata causes a decrease in their number from 10-15 to 2-3 in each conjugating pair. The most common types of chromosome shape in metaphase are rings and rods. In some cases i.e. *A. nutans*, pairing in meiosis is incomplete in certain chromosome pairs (Levan, 1931). During meiosis in the polyploid forms, associations of chromosomes are common. Appearance of trivalents in *A. nutans* was reported (Levan, 1935). Observed trivalents had different forms, i.e V-shaped trivalents with one rod on each arm or with two rods on one arm. In anaphase I trivalents are distributed randomly, usually twelve to the pole, however, formation 8 to one pole and 16 to other was also observed (Levan, 1931). Occurrence of lagger chromosomes developing micronuclei, as well as formation of unreduced gametes was reported in triploids (Ved Brat, 1967). Quadrivalents were formed in meiosis I in the PMC of *A. schoenoprasum* and *A. porrum*. The most frequent types of quadrivalents were rings and chains (Levan, 1935). In polyploid *A. oleraceum* (2n=24, 32, 40) elimination of chromosomes during meiosis was observed, thus pollen grains with 13, 15, 17, and other number of chromosomes were noted (Levan, 1931).

2.1 *A. cepa*

Allium cepa L. belongs to the genus *Allium*, subgenus *Rhizirideum*, section *Cepa* (Mill.) (Fritsch & Friesen, 2002). It is a diploid (2n=2x=16) and possesses one of the largest genomes among cultivated plants (Havey, 2002). The nuclear genome of onion contains 35,8 pg per 2C (Labani & Elkington, 1987), which is reflected by very large chromosomes. Onion has a complement of eight pairs of metacentric and submetacentric chromosomes including one set of terminal satellite chromosomes. Meiosis in onion is regular with eight bivalents (R.N. Jones, 1990). Chiasmata during metaphase I in PMC in onion are distributed at random in each bivalent and by the mid-metaphase they become terminal or sub-terminal and either rod or ring bivalents are formed (Fig. 1 & Fig. 2) (Emsweller & H.A. Jones, 1945; Levan, 1936). The cross-over points are therefore located mainly in distal and interstitial regions of the chromosome arms (Koul & Gohil, 1970; Levan, 1933). There is no information about localized chiasmata in onion (Emsweller & H.A. Jones, 1945).

2.2 Hybrids with introgression from *A. fistulosum*

Allium fistulosum L. (Japanese bunching onion, Welsh onion) is a diploid species (2n=2x=16). *A. fistulosum*, similarly to onion, belongs to section *Cepa* of the genus *Allium* (R.N. Jones, 1990). Nuclear DNA content is 26,3 pg (2C), what is about 20% less than in onion (R.N. Jones & Rees, 1968).

During meiosis I in *A. fistulosum* eight bivalents are formed. Bivalents are held together by two chiasmata, one on each side of the centromere, thus in the metaphase plate they appear like a cross (Fig. 2). Chiasmata in the bivalent are usually localized proximally, adjacent to the centromere, however, occasional formation of more interstitial and randomized chiasmata were observed (R.N. Jones, 1990; Levan, 1933; Maeda, 1937).

Progenies from generative hybridization of onion with *A. fistulosum* have been studied most extensively among all interspecific crosses in the genus *Allium*. Those two species can be easily hybridized in the greenhouse and the success rate can be enhanced by the use of embryo-rescue in tissue cultures (Doležel et al., 1980). First hybrids between *A. cepa* and *A. fistulosum* were reported in 1935 by Emsweller and Jones. Meiotic studies showed that in early prophase I the hybrid chromosomes appear to be single threads, like in the parents. Abnormalities in the cell division start at late prophase, what was confirmed by Maeda (1937). In pachytene, single threads are doubled and chiasmata are visible between some conjugating partners, whereas other chromosomes probably just overlapped each other at entire length of the arms. When the bivalents were formed by chromosomes of different length, they sometimes separated in association long plus long chromatid, as well as long plus short chromatid. Emsweller and Jones (1935b) observed extension of long arm of conjugated chromosomes well beyond the end of the other. That condition depends on the arrangement of chiasmata at the ends of a bivalent (Emsweller & H.A. Jones, 1935a). In diakinesis, chromosomes were usually well-separated and their configuration could be easily determined. In *A. cepa*, chiasmata are predominantly terminal, in *A. fistulosum* chiasmata are localized near the centromere region (Fig. 1). The hybrids have chiasmata localized randomly (Emsweller & H.A. Jones, 1945; Maeda, 1937). Hybrids had high frequency of complete bivalent pairing, but unpaired chromosomes were also observed (Emsweller & H.A. Jones, 1935a, 1938). In anaphase I, when the separations of chromosomes occur, occasional chromosome bridges were present (Emsweller & H.A. Jones, 1938).

Emsweller and Jones (1935a, 1935b, 1938) as well as Peffley (1986) and Maeda (1937) reported regular bivalentization (Fig. 2) in about 70% of tested cells, whereas Levan (1941) reported regular bivalent formation in only 2% of the tested PMC in his *A. cepa* x *A. fistulosum* hybrids. Early stages of meiosis in that plant material showed a number of unpaired threads, the maximal pairing was found before diplotene. Few pachytene chromosomes had threads longer than their pairing partner, what in consequence lead to occurrence of loop shape formations. Loops were localized on one side of the pairing complex, but in some cases the longer thread folded and paired with itself. Ring (Fig. 2.) and chromosome configurations with three chiasmata were also found (Levan, 1935). Inter-chromosomal differences, resulting in pairing of more than two chromosomes were observed. Levan (1941) observed frequent trivalents (14%), forming rings in metaphase. Seldom quadrivalents, pentavalents, hexavalents, were reported (Emsweller & H.A. Jones, 1935b, 1938; Levan, 1936; Peffley, 1986). However, Maeda (1937) reported absence of associations higher than bivalents in his hybrids. Abnormalities observed by Peffley (1986) in anaphase I included bridge formation in 28% and lagging chromosomes in 10% of the cells. Deficiencies, translocations, and inversions were reported (Emsweller & H.A. Jones, 1938; Levan, 1935, 1941; Peffley, 1986). Presence of heteromorphic pairing in F_1 interspecific hybrids was observed (Emsewller & H.A. Jones, 1935b; Maeda, 1937). Peffley (1986) suggested that heteromprphic bivalents were the consequence of pairing between chromosomes with translocation or inversion at the centromeric region.

The above-described discrepancies in meiotic events in F1 A. cepa x A. fistulosum hybrids may be explained by a different origin of A. cepa and A. fistulosum used in the discussed studies. Maeda and Emsweller and Jones used the same variety of A. cepa (Yellow Danvers) and A. fistulosum in Japanese form (Nebuka, Hidanegi), whereas Levan mentioned the use of a commercial onion variety 'Braunschweiger' and as a source of fistulosum points one of the European botanical gardens. Peffley (1986) used onion cv. 'Yellow Grano' and ten different accessions of A. fistulosum supplied by International Plant Breeders. Thus, it can be concluded that the cytological behavior during meiosis in F1 interspecific hybrids seems to be accession-specific. Moreover, chromosomal variants (multivalents) and changes in the chromosome structure (bridges) offer an explanation of the low pollen fertility of the F_1 hybrids (H.A. Jones & Mann, 1963; Levan, 1941; Peffley, 1986).

Using GISH (Genomic In Situ Hybridization) Stevenson et al. (1998) detected reciprocal crossover events in F_1 interspecific hybrids. The frequencies of crossovers detected as label exchanges in anaphase I chromosomes were about 20% higher than metaphase I chiasma frequencies. Since the synaptonemal complex (SC) is responsible for transmission of the interference, discontinuities in SC may cause that crossover occur in much closer proximity than they normally would with uninterrupted SCs (Sybenga, 1996). Albini and Jones (1990) reported incomplete synapsis in the centromeric region and other irregularities in the SC in the A. cepa x A. fistulosum hybrids. The failure of synapsis might explain differences in chiasmata frequencies reported by Stevenson et al. (1998). Additionally, Albini and Jones (1990) suggested, that DNA differences between A. cepa and A. fistulosum chromosomes were localized in the centromeric region, which prevented regular synapsis or progression of synapsis and possible proximal chiasma formation.

The F_1 hybrids between A. cepa and A. fistulosum can be easily produced, but have low pollen fertility, often not exceeding 10% (H.A. Jones & Mann, 1963; Levan, 1941; Maeda, 1937; Peffley, 1986; van der Valk et al., 1991b). Additionally Emsweller and Jones (1938) reported occurrence of egg sterility of the F_1 hybrids. Successful production of F_2 progenies was reported by Levan (1941). The F_2 plants were exclusively polyploids (tri- and tetraploids). Analysis of meiosis showed frequent disturbances, such as chromatin bridges and micronuclei in dyads and tetrads. Random distribution of chiasmata was dominant.

Cytological analysis of tetraploid forms revealed that they were amphidiploids built up from 2 *cepa* and 2 *fistulosum* genomes. Morphology of the tetraploids was intermediate between the parents, while their pollen fertility varied. One tetraploid was sterile, the other had moderate (50%) pollen fertility. In the triploid form, the frequency of trivalents was high (80-90%). In about 33% of tested cells micronuclei and chromosomal bridges were observed. Pollen fertility was low; not exceeding 10% (Levan, 1936, 1941).

Levan's (1941) attempts to backcross (BC) F_1 hybrid *A. cepa* x *A. fistulosum* to *cepa* were unsuccessful. The studies of van der Valk et al. (1991a) showed that difficulties in this type of cross were due to pre-fertilization barrier, as the growth of onion pollen tubes in the style of the hybrid was disturbed. Maeda (1937) backcrossed F_1 to *A. fistulosum* and obtained viable seeds. Emsweller and Jones (1945) obtained several BC progenies, but only when the hybrid was used as the pollen source.

In BC_1, the most common meiotic configuration were eight bivalents and eight univalents. Both randomized and localized chiasmata were observed (Emsweller & H.A. Jones, 1935b). Multivalents (quadrivalents, pentavalents) were observed in about 40% of the meiocytes. A ring univalent, lagging chromosomes, and chromosomal bridges were also observed (Peffley & Mangum, 1990). Analysis of the structural differences of chromosomes revealed presence of at least three paracentric inversions and one translocation, producing changes in gene order (Ulloa et al., 1994). Emsweller and Jones (1945), using F_1 as a pollen source, backcrossed them to both *A. cepa* and *A. fistulosum*. On average 44% of PMC in population backcrossed to *A. cepa* showed regular chromosome pairing, additionally those plants segregated with regard to male fertility; some individuals after self-pollination yielded in few seeds, and some were complete sterile. The first backcross to *A. fistulosum* possessed both randomized and localized chiasmata, however the second backcross had a high frequency of localized chiasmata (Fig. 1). This increase of localized chiasmata in subsequent backcrosses was correlated with pollen fertility.

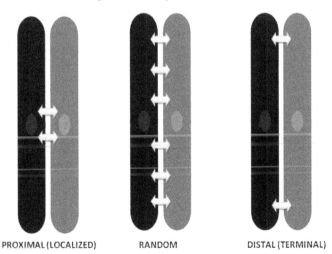

PROXIMAL (LOCALIZED)   RANDOM   DISTAL (TERMINAL)

Fig. 1. Diagram showing the types of chiasmata distribution in the interspecific F1 hybrids within edible alliums.
Arrows indicating possible position of chiasmata between two homologous chromosomes in bivalent.

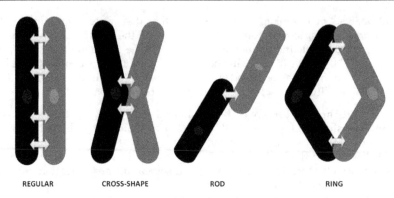

REGULAR   CROSS-SHAPE   ROD   RING

Fig. 2. Diagram showing configurations of bivalents observed in metaphase I of interspecific F1 hybrids within edible alliums.
Arrows indicating position of chiasmata between two homologous chromosomes in bivalent.

The fertile BC$_2$ plants had localized chiasmata and their morphology was similar to *A. fistulosum*, while sterile plants had mostly randomized chiasmata and morphologically were in the *A. cepa* type. Hou and Peffley (2000) examined 16 BC$_3$ plants and observed very high pollen fertility (90-92%) of the plants with eight bivalents at metaphase I. Those plants had low percentage of chromosomal aberrations such as bridges, laggards, or micronuclei. One male sterile plant had recombinant chromosomes. Thirteen among sixteen chromosomes in PMCs of the sterile plant gave strong hybridization signals at telomeres when *A. fistulosum* was used as a genomic DNA. Similar GISH patterns were reported by Khrustaleva and Kik (1998), suggesting differences in heterochromatin distribution at the telomeres between two parental species.

Khrustaleva and Kik (1998) introgressed genes from *A. fistulosum* to *A. cepa* using *A. roylei* as a bridge in crossing. Multi-colour GISH was performed to analyze genome organization of the hybrid. Results showed 8 *cepa* chromosomes, 1 metacentric chromosome from *roylei*, and 7 recombinant chromosomes. Recombination between *A. roylei* and *A. fistulosum* took place in distal and interdistal regions of hybrid chromosomes, centromeric regions originated from *roylei*. Meiosis in PMC of the first generation bridge-cross hybrid [*A. cepa* x (*A. fistulosum* x *A. roylei*)] was rather regular, about 53% of tested cells possessed 8 ring bivalents (Fig. 2), however occasional presence of univalents was noted. No multivalents were observed. Some of the bivalents were heteromorphic, indicating homologous pairing and chiasmata formation during prophase I. Pollen viability of the hybrid was about 50-60%. Detailed analysis of meiosis in the second generation hybrid (*A. cepa* x first generation bridge cross) showed that recombination between three genomes was frequent and that chiasmata were randomly distributed (Fig. 1) (Khrustaleva & Kik, 2000). Keeping in mind that chiasmata are located adjacent to the centromere in *A. fistulosum* and randomly distributed in *A. cepa* and *A. roylei* (Emsweller and H.A. Jones, 1935b; Levan, 1933, de Vries et al., 1992) in the bridge-cross hybrid random chiasmata prevailed over the *fistulosum* type. Translocations of *roylei-fistulosum* segment and deficiencies (absence of *roylei* segment) were also reported (Khrustaleva & Kik, 2000). Pollen fertility of the second generation bridge-cross hybrid was variable, what might be caused by the use of CMS onions as the female parent and/or the lack of *roylei* chromosome segments containing male sexual reproduction genes (Khrustaleva & Kik, 2000; Ulloa et al., 1995).

Monosomic addition lines of *fistulosum – cepa* (Shigyo et al., 1998) and *cepa – fistulosum* (Hang et al., 2004; Peffley et al., 1985) were reported. These lines are valuable to study genome organization in *A. cepa* and *A. fistulosum*.

2.3 Hybrids with introgression from *A. ampeloprasum* spp. *porrum*

Leek is a tetraploid species (2n=4x=32). Nuclear DNA content is 52,7 pg per 2C (Bennett & Leitch, 2010). Leek belongs to the genus *Allium*, subgenus *Allium*, section *Allium* (Hirschegger et al., 2010).

Meiosis in the leek was first described by Levan in 1940, showing almost complete localization of chiasmata at prophase I. Khazanehdhari et al. (1995) reported frequent formation of quadrivalents at this stage. Further studies showed that those quadrivalents were resolved in to bivalents in metaphase I (G.H. Jones et al., 1996); however, occasional persistence of quadrivalents in this stage was also observed (Koul & Gohil, 1970; Levan, 1940). Formation of bivalents in metaphase I was due to proximal localization of chiasmata in leek (Stack & Roelofs, 1996). Usually, four chromosomes in tetrasome are clustered by chiasmata located immediately adjacent to and on either side of the median or sumbedian centromere. As pairing partner switches are unlikely to occur between two proximal chiasmata, thus quadrivalent fall apart into two bivalents having a characteristic cross-shape (Fig 2.) (G.H. Jones et al., 1996; Khazanehdhari et al., 1995). Occurrence of univalents in the PMC of the leek was also reported (G.H. Jones et al., 1996; Khazanehdhari & G.H. Jones, 1997).

Interspecific hybrids between *Allium cepa* and *A. ampeloprasum* were generated with the aim of introduction of the S-cytoplasm from onion into leek. Few attempts of sexual hybridization of onion and leek failed, suggesting that this strategy to create F1 progeny may be difficult (Doležel et al., 1980; Currah, 1986). Peterka et al. (1997) used *in vitro* culture to rescue the F_1 embryos at the early stages of development (7-14 days after pollination); as a result they obtained seven hybrid plants, which were triploids with 24 chromosomes. Authors reported presence of three chromosomes with satellites and two with intercalary pseudosatellites. Onion carries two chromosomes with satellites (Kalkmam, 1984; Taylor, 1925), leek has four, and additionally leek has four chromosomes with intercalary pseudosatellites (Murin, 1964). GISH with onion DNA as a probe, showed hybridization of eight chromosomes of the hybrid plant, the remaining 16 did not hybridized, which enabled identification of onion and leek chromosomes. Further studies of meiosis in the interspecific hybrids showed presence of eight leek bivalents and eight onion univalents in the prophase I (Peterka et al., 2002). The leek bivalents in the hybrids had localized chiasmata (Fig. 1). During late metaphase and early anaphase, in some of the observed cell the onion univalents were arrested at the periphery of the nucleus, while the leek chromatids moved to the poles. The onion chromatids, due to their retarded movement, were frequently excluded from the daughter nuclei in the form of micronuclei observable in the dyad or tetrad stage (Schrader et al., 2000).

Peterka et al. (2002) investigated chromosome composition of the backcross progenies. Their BC_1 plants always possessed eight onion chromosomes and from 30 to 33 leek chromosomes. Only in one tested hybrid a recombinant chromosome was identified. In the BC_2 and BC_3 decreased transmission rate of univalent onion chromosomes were observed, resulting in the production of alloplasmic leek plants in the third backcross.

Buiteveld et al. (1998a, 1998b) reported the symmetric fusion of protoplasts isolated from suspension cultures of *A. ampeloprasum* and leaf mesophyll protoplasts of onion, resulting in somatic hybridization of onion and leek. Analysis of chromosome composition of somatic hybrids was performed using GISH. All obtained hybrids were identified as hexaploids. In eight tested hybrids, chromosome number varied from 41 to 45. Authors found differences among hybrids with regard to the number of parental chromosomes. Some hybrids carried eight onion and less than 32 leek chromosomes, other possessed 12 onion and 30 leek chromosomes. Flow cytometry measurements showed that the suspension culture used for the fusion was of a mixture of aneuploid (hypotetraploid) and normal tetraploid cells of leek. Since aneuploid leek protoplasts were used for the fusion, it is possible that the hybrids resulted from fusions with leek protoplasts with different chromosome numbers, which might explain the differences in the number of leek chromosomes in the hybrids. Loss of the onion chromosomes may be a consequence of spontaneous chromosome elimination after the fusion. Moreover, recombinant chromosomes in cells of leek and onion somatic hybrids were also reported. Two of those chromosomes were of a reciprocal translocation type and one was an interstitial translocation type (Buiteveld, 1998a).

2.4 Hybrids with introgression from *A. sativum*

A. sativum L. is a diploid with (2n=2x=16) with nuclear DNA content 32,5 pg per 2C (Bennett & Leitch, 2010). The DNA content of garlic is the most similar to onion among the discussed species. Garlic belongs to the genus *Allium*, subgenus *Allium*, section *Allium* (Hirschegger et al., 2006; Keller et al., 1996).

The majority of garlic cultivars are sterile, which preclude to use them as a partner in sexual crossing, however occurrence of fertile garlic plants was also reported (Hong et al., 1997; Kik, 2002). Flowering in some garlic clones is often associated with presence of bulbils formed in the place of aborted flowers. Bulbils are used as a source material for vegetative reproduction (R.N. Jones, 1990; Kik, 2002).

To study meiosis in the garlic, Konvička and Levan (1972) used two fertile clones named OH and LH. The clone OH formed medium size bulbils and normal flowers in the inflorescence. The clone LH had an abundance of flowers together with many small bulbils, moreover this clone formed original chromosome ring at meiosis (amphibivalent) (Levan, 1936). Majority of tested meiocytes in the clone OH formed eight bivalents at meiosis I; however presence of spherical chromatic bodies outside the spindle was observed. Produced pollen grains differ in size. The first meiotic division in the clone LH proceeded regularly, but two out of four bivalents formed an ring consisting of four chromosomes. This structure could be observed from diplotene through diakinesis and metaphase I. In metaphase I authors observed ring configurations in the majority of the cells; only in one of them an open ring was present. The formation of rings caused occurrence of lagging chromosomes observable during the first anaphase. The second division was usually normal, but in few cells chromatin body outside the spindle was present. Two satellite chromosome pairs were noted in cells of both tested clones. The clones had their own characteristic pattern of meiotic abnormalities, what is in contrary to results presented by Koul and Gohil (1970) where in three tested fertile clones meiosis was completely regular. Differences in chromosome length, positions of NOR or arm ratios were variable among garlic clones (R.N. Jones, 1990). Such differences may partially develop after the loss of sexual propagation and changes occurring in the somatic cells of vegetatively propagated garlic.

Allium cepa and *A. sativum* for a long time have not been hybridized sexually because of the narrow pool of fertile garlics and a large genetic distance (Kik, 2002; Keller et al., 1996). The first interspecific hybrids between onion and garlic were reported by Ohsumi et al. in 1993. Authors performed reciprocal crosses of onion and garlic, but embryos were formed only when onion was the female parent. Embryos were rescued in tissue culture. Regenerated hybrids possessed 16 chromosomes in somatic cells, with clearly distinguishable two satellite chromosomes. Since onion cv. 'Sapporoki' used in the study did not possess satellite chromosomes, those chromosomes were inherited from *A. sativum*. Chromosome elimination was not observed in the study. The pollen viability of the obtained hybrid vas very low (2%).

Worth mention is work of Yanagino et al. (2003) with aim of increasing fertility in garlic through sexual hybridization with leek. Success of this cross was accelerated, since both leek and garlic belong to the subgenus *Allium*. Leek was used as the female partner. Interspecific hybrids were obtained with the use of the embryo-rescue technique. Hybrids were triploids and possessed 24 chromosomes (16 from onion and 8 from garlic), however meiosis in PMC was not examined. Obtained hybrids were almost completely sterile.

Somatic hybrids of onion and sterile garlic were reported by Yamashita et al. (2002). Onion and garlic have sixteen chromosomes, however obtained hybrids were classified as aneuploids with total number of chromosomes 40 or 41. In the tested somatic cells of the hybrids authors noted presence of two subtelocentric chromosomes and three intercalary satellite chromosomes, inherited from both parents. GISH analysis revealed 17 garlic and 20 or 21 onion chromosomes and three chromosomes consisting of chromosomal regions from both parents. Presence of chimeric chromosomes in the somatic hybrids originated from translocation between chromosomes or chromosome fusions.

3. Conclusions and future directions

Pre- fertilization barriers in distant crosses, such as failure in pollen germination, slow pollen tube growth, or foreign pollen tube arrest in the style have been reported in different species (Kiełkowska & Adamus, 2006; Manickam & Sarkar, 1999b) including alliums (Gonzalez & Ford-Lloyd, 1987; Ohsumi et al., 1993; van der Valk et al., 1991a). Post-fertilization barriers are related to abnormalities in the development of the zygote, absence or abnormal development of the endosperm which causes embryo starvation and abortion (Zenkteller, 1990). Within the discussed species, the embryos developed regularly in the hybrids with introgression from *A. fistulosum*, or had to be rescued on the early stages of the development in hybrids with the leek and garlic introgression. Cytological studies showed reproductive abnormalities in F1 hybrids and their progenies. Irregularities of chromosome pairing and occurrence of a range of structural changes in meiotic chromosomes resulted in unbalanced chromosome complements in the gametes, often causing lowered fertility or even sterility of the F_1 hybrids (Khush & Brar, 1992; Peffley, 1986; van der Valk et al., 1991a).

A. fistulosum carries resistance genes for fungal, bacterial, and viral diseases (Rabinowitch, 1997). Moreover *fistulosum* has several agronomically important traits such as high dry-matter content, winter-hardiness, high pungency and earlier flowering as compared to onion (van der Meer & van Bennekom, 1978). Hence, hybrids between onion and *A. fistulosum* as well as several backcross progenies were deeply studied, with *A. cepa* as either the female parent (Emsweller & H.A. Jones, 1935a, 1935b; Levan, 1941; Van der Meer and Van Benekom, 1978) or pollen parent (Peffley, 1986; Peters et al., 1984; van der Valk et al. 1991a). Cytological studies have shown that chromosome pairing is highly variable in the F_1 hybrid depending on the parental combination (Emsweller & H.A. Jones, 1935a, 1953b; Peffley, 1986; Ulloa et al., 1994). Presence of heteromorphic pairing was reported in each of the discussed studies, showing evidence for recombination of genetic material between homologous chromosomes in the F1 hybrid between A. cepa and A. fistulosum (Emsweller & H.A. Jones, 1935a, 1935b; Peffley & Mangum, 1990; van der Meer & Van Benekom, 1978), and when the hybrid was backcrossed to A. cepa (Hou & Peffley, 2000). Khrustaleva and Kik (1998, 2000) showed that gene exchange is possible even in the three-way (cepa x (fistulosum x roylei)) hybrid, resulting in partially fertile progenies. Nucleo-cytoplasmic incompatibility interactions reported between these two species may also reduce fertility of F_1 hybrids (Ulloa et al., 1994).

Although pollen sterility of hybrids is a major barrier to gene introgression, occurrence of fertile F1 hybrids able to produce further progenies and introduce them into the breeding process, was also reported. Jones and Clarke (1942) reported obtaining of amphidiploid arose by spontaneous doubling from unreduced gamete of F_1 A. cepa x A. fistulosum hybrid. At metaphase I of the hybrid, there were 16 bivalents with localized and non-localized chiasmata and meiosis was regular. The obtained plant was fertile, vigorous, and resistant to several diseases. This amphidiploid is known as 'Beltsville Bunching' and was released in the United States in 1950, and sold on the market as green bunching onion (H.A. Jones & Mann, 1963). Many years of research resulted in the release of the several other hybrids grown as vegetatively propagated onions in fistulosum type or sexually reproduced bulb onions resistant to pink root (Phoma terrestris) (for review see Kik, 2002).

In the leek, the main problem concerning sexual hybridization lies in the high (98%) occurrence of chiasmata localized near the centromere (Levan, 1940). Such localized chiasmata prevent formation of multivalents during meiosis (metaphase I) and prevents recombination on the distal ends of chromosomes (Khazanehdari & G.H. Jones 1997). The localization of chiasmata in this near-centromeric section prevents association of more than two chromosomes during leek meiosis, which may be beneficial for its capacity to form fertile gametes. However, if the pairing behavior in metaphase I is different between sexually hybridized components that may lead to infertility of the hybrids (Brewster, 1994). Peterka et al. (2002) reported complete sterility of onion x leek hybrids.

Morphology and hybrid status of progenies from a cross between A. cepa x A. sativum were reported (Ohsumi et al., 1993), however, meiotic chromosomes behavior in the hybrids as well as their pollen fertility have not been deeply investigated. Broadening genetic diversity in garlic is very desired. It can be achieved with the use of protoplast cultures, which offers a large spectrum of possibilities with regard to selection of components for the fusion. Sexual hybridization is rather difficult in garlic, due to high seed sterility of the clones. However, utilization of wild relatives and search for novel fertile accessions like the one found in South America (Hirschegger et al., 2006) is very important for increasing available germplasm pool.

During the years of studies researchers attempted to develop techniques helping to overcome barriers between incompatible species for successful production of hybrid seeds.

Worthy mentioning is bridge crossing (Dionne, 1963; McCoy & Echt 1993), use of exogenous plant growth regulators, i.e. gibberelic acid, dichloro-acetic acid (Brock, 1954; Manickam & Sarkar, 1999a), mixed pollen technique (Asano & Myodo, 1977), use of the 'mentor' pollen (Sastri & Shivanna, 1976; Settler, 1968), *in vitro* fertilization (De Verna et al., 1987; Zenkteller et al., 2005), embryo rescue (Sukno et al., 1999; Williams et al., 1982), and somatic cell hybridization (Kirti et al., 1991; Smith, 1976).

Embryo rescue promotes the development of an immature or weak embryo into a viable plant. The most commonly used procedure is careful excision of hybrid embryos and placing directly onto the culture medium. Sometimes when there are difficulties with embryo excision or embryos are very small, whole ovules (ovule culture, *in ovolo*) or ovaries are put to the culture (Bridgen, 1994). Embryo rescue has been widely used for producing hybrid plants in onions. *Alliums* hybrid embryos were usually cultured on a phytohormone-free medium (Amagai et al., 1995; Gonzalez & Ford-Lloyd, 1987; Nomura & Makara, 1993; Umehara et al., 2006). Whole ovary culture supported maturation of viable seeds in hybrids of *A. cepa* x *A. sphaerocephalon* (Bino et al., 1989) or *A. fistulosum* x *A. schoenoprasum* (Umehara et al., 2007). Keller et al. (1996) adopted ovary culture and obtained many hybrids between onion and other distant species. This technique was used in the crosses between onion and *A. fistulosum* (Doležel et al., 1980; van der Valk, 1991a), onion and leek (Peterka et al., 1997), onion and garlic (Ohsumi et al., 1993), rakkyo (*A. chinense*) and *A. fistulosum* L. (Nomura et al., 1994; Nomura & Makara, 1993), Welsh onion and chives (*A. schoenoprasum*) (Umehara et al., 2006) and many others (Gonzalez & Ford-Lloyd, 1987; Nomura et al., 2002; Nomura & Oosawa, 1990; Yanagino et al., 2003). Application of protoplast cultures for obtaining interspecific crosses in edible alliums (Buiteveld et al., 1998a, 1998b; Shimonaka et al., 2002; Yamashita et al., 2002) was also reported. Genomic *in situ* hybridization enabled the identification of genomes in the interspecific and intergeneric hybrids together with direct detection of genetic recombination (Friesen et al., 1997; Schwarzacher et al., 1989; Thomas et al., 1994). It seems that studies on the interspecific crosses in alliums employed a wide range of different biotechnological and cytogenetical tools to facilitate obtainment and characterization of hybrid, however one issue still remains unsolved and problematic. Many authors pointed low pollen fertility of generated hybrids what in consequence significantly decreased the number of obtained seeds and narrowed the pool of accessions for cytogenetic studies (H.A. Jones & Mann, 1963; Peffley, 1986; van der Valk et al., 1991a, 1991b). Most of the hybrids from crosses of *A. cepa* with *A. fistulosum* (Doležel et al., 1980; Emsweller & H.A. Jones 1935a, 1935b; Maeda, 1937), leek (Doležel et al, 1980), garlic (Ohsumi et al., 1993), as well as with *A. galanthum* (Kiełkowska & Adamus, 2010), *A. oschaninii* or *A. pskemense* (McCollum, 1971) were reported as sterile.

Possibly, restoration of fertility of interspecific hybrids can be achieved by chromosome doubling (R.N. Jones, 1983). Chromosomes are doubled by treating the plant tissue with antimitotic agents such as colchicine (Blakslee, 1939; Blakslee & Avery, 1937). McCollum (1980) reported fertile amphidiploid of *A. cepa* x *A. galanthum* obtained by colchicine treatment. Autotetraploids in *A. cepa* and *A. fistulosum* were obtained by soaking germinating seeds in aqueous colchicine solution, but the recovery of tetraploids was low (2%) (Toole & Clarke, 1994). Song et al. (1997) treated calli form F_1 hybrids of *A. fistulosum* x *A. cepa* with 0.1 and 0.2% colchicine and regenerated tetraploids. It seems that the potential is promising and the recovery of fertile plants from colchicine-treated F_1 hybrids has been reported in other species (Eigsti & Dustin 1955; Orton & Steidl, 1980). The optimum concentration and duration of treatment has to be determined empirically as it strongly affects the success in doubling and the survival of the treated plant material (Eikenberry, 1994; Hansen & Andersen 1996; Klima et al., 2008; Wan et al., 1989). In the literature concerning doubled haploids, authors mention use of trifluraline, oryzaline or amiprophos methyl (Hansen and Andersen, 1996; Zhao & Simmonds, 1995) which are considered less toxic and may be beneficial for fertility restoration in onion interspecific hybrids.

The state of knowledge about meiosis in F_1 hybrids and further generations from crossings of onions with A. fistulosum was deeply investigated, however meiotic studies of hybrids between onion and garlic and leek is limited. The change of the status is needed, because the understanding of meiotic process in crucial for further research on the reproduction, fertility, genetics and breeding of discussed species.

4. References

Albini , S.M. & Jones G.H. (1990) Synaptotemal complex spreading in *Allium cepa* and *Allium fistulosum*. III. The Hybrid. *Genome*, Vol. 33, pp. 854-866

Amagai, M.; Ohashi K. & Kimura S. (1995). Breeding of interspecies hybrid between *Allium fistulosum* and *A. tuberosum* by embryo culture. *Bull Tochigi Agr Exp Stn*, Vol. 43, pp. 87–94

Asano, Y. & Myodo H. (1977). Studies on crosses between distantly related species of lilies for the intrastyllar pollination technique. *J Jpn Soc of Hort Sci*, Vol. 46, pp. 59-65

Bennett, M.D. & Leitch, I.J. (2010). Plant DNA C-values Database (release 5.0, December 2010) http://data.kew.org/cvalues/

Bino, R.J.; Janssen M.G.; Franken, J. & de Vries, J.N. (1989). Enhanced seed development in the interspecific cross *Allium cepa* x *A. sphaerocephalon* through ovary culture. *Plant Cell Tiss and Org Cult*, Vol. 16, pp. 135-142

Blakeslee, A. & Avery A. (1937). Methods of inducing doubling of chromosomes in plants by treatment with colchicine. *J Hered*, Vol. 28, pp. 393–411

Blakeslee, A. (1939). The present and potential service of chemistry to plant breeding. *Am J Bot*, Vol. 26, pp. 163–172

Bougoud, S.M. & Plowman A.B. (1996). The inheritance of B chromosomes in *Allium schoenoprasum* L. *Chromosome Res*, Vol. 4, pp. 151-158

Bougourd, S.M. & Parker J.S. (1979). The B-chromosome system of *Allium schoenoprasum*. *Chromosoma*, Vol. 75, pp. 369-383

Brewster, J.L. (1994). The genetics and Plant breeding of *Allium* crops. In: *Onions and other vegetable alliums*. Brewster J.L., pp. 41-61, CAB International, ISBN 0581995101,UK

Bridgen, M.P. (1994). A review of plant embryo culture. *HortScience*, Vol. 29, pp. 1243–1246

Brock, R.D. (1954). Hormone induced pear-apple hybrids. *Heredity*, Vol. 8, pp. 421-429

Buiteveld, J.; Kassies W.; Geels R.; van Lookeren Campagne M.M., Jacobsen E. & Creemers-Molenaar J. (1998b). Biased chloroplast and mitochondrial transmission in somatic hybrids of *Allium ampeloprasum* L. and *Allium cepa* L. *Plant Sci*, Vol. 131, pp. 219-228

Buiteveld, J.; Suo Y.; van Lookeren Campagne M.M. & Creemers-Molenaar J. (1998a). Production and characterization of somatic hybrid plants between leek (*Allium ampeloprasum* L.) and onion (*Allium cepa* L.). *Theor Appl Genet*, Vol. 96, pp. 765-775

Currah, L. (1986). Leek breeding: a review. *J Hortic Sci*, Vol. 61, pp. 407-415

De Verna, J.W.; Myers J.R. & Collins G.B. (1987). Bypassing prefertilization barriers to hybridization in *Nicotiana* using *in vitro* pollination and fertilization. *Theor Appl Genet*, Vol. 73, pp. 665-671

de Vries, J.N.; Wietsma W.A.; & Appels M. (1992). Direct and *Allium roylei* mediated transfer of *A. fistulosum* genes to onion. *Allium Improvement Newsletter*, Vol. 2, pp. 9-10

Dionne, L.A. (1963). Studies on the use of *Solanum acaule* as a bridge between *Solanum tuberosum* and species in the *Bulbocastana, Cardiophylla* and *Pinnatisecta. Euphytica*, Vol. 12, pp. 263-269

Doležel, J.; Nowak F.J. & Lužny J. (1980). Embryo development and *in vitro* culture of *Allium cepa* and its interspecific hybrids. *Z. Pflanzenzüchtg*, Vol. 85, pp. 177-184

Eigsti, D.I. & Dustin P. Jr (1955). Spindle and cytoplasm. In: *Colchicine in agriculture, medicine, biology and chemistry*, pp. 65-139, The Iowa State University College Press, ISBN 9780598806932, Ames

Eikenberry, E. (1994). Chromosome doubling of microspore- derived canola using trifluralin. *Cruciferae Newsletter*, Vol. 16, pp. 51–52

Emsweller, S.L. & Jones H.A. (1935a). An interspecific hybrid in *Allium. Hilgardia* Vol. 9, pp. 265-273

Emsweller, S.L. & Jones H.A. (1935b). Meiosis in *A. fistulosum, A. cepa* and their F1 hybrid. *Hilgardia* Vol. 9, pp. 277-294

Emsweller, S.L. & Jones H.A. (1938). Crossing-over, fragmentation, and formation of new chromosomes in an *Allium* species hybrid. *Botanical Gazette*, Vol. 99, pp. 729-772

Emsweller, S.L. & Jones H.A. (1945). Further studies on the chiasmata of the *Allium cepa* x *A. fistulosum* hybrid and its derivatives. *American Journal of Botany*, Vol. 32, pp. 370-379

Etoh, T. & Simon P.W. (2002). Diversity, fertility and seed production of garlic. In: *Allium Crop Science: recent advances*, Rabinowitch H.D. & Currah L., pp. 101-117, CABI Publishing, ISBN 0851995101, UK

Friesen, N.; Fritsch R. & Bachman K. (1997). Hybrid origin of some ornamentals of Allium subgenus *Melanocrommyum* verified with GISH and RAPD. *Theor Appl Genet*, Vol. 95, pp. 1229-1238

Fritsch, R.M. & Friesen N. (2002). Evolution, domestication and taxonomy. In: *Allium Crop Science: recent advances*. Rabinowitch H.D. & Currah L., pp. 5-30, CABI Publishing, ISBN 0851995101, UK

Fuchs, J.; Brandes A. & Schubert I. (1995). Telomere sequences localization and karyotype evolution in higher plants. *Plant systematic and Evolution*, Vol. 196, pp. 227-241

Gonzalez, L.G. & Ford-Lloyd B.V. (1987). Facilitation of wide crossing through embryo rescue and pollen storage in interspecific hybridization of cultured *Allium* species. *Plant Breed*, Vol. 98, pp. 318–322

Hang, T.T.M.; Shigyo M.; Yamaguchi N. & Tashiro Y. (2004). Production and characterization of alien chromosome addition in shallot (*Allium cepa* L. Aggregatum group) carrying extra chromosome(s) of Japanese bunching onion (*A. fistulosum* L.). *Genes Genet Syst*, Vol. 79, pp. 263-269

Hansen, N.J.P. & Andersen S.B. (1996). *In vitro* chromosome doubling potential of colchicine, oryzalin, trifluralin, and APM in *Brassica napus* microspore culture. *Euphytica*, Vol. 88, pp. 156–164

Havey, M.J. (2002). Genome organization in *Allium*. In: *Allium Crop Science: recent advances*. Rabinowitch H.D. & Currah L., pp. 59-79, CABI Publishing, ISBN 0851995101, UK

Hirschegger, P.; Galmarini C. & Bohanec B. (2006). Characterization of novel form of fertile great headed garlic (*Allium* sp.). *Plant Breed*, Vol. 125, pp. 635-637

Hirschegger, P.; Jakše J.; Trontelj P. & Bohanec B. (2010). Origins of *Allium ampeloprasum* horticultural groups and a molecular phylogeny of the section *Allium* (*Allium: Alliaceae*). *Molecular Phylogenetics and Evolution*, Vol. 54, pp. 488-479

Hong, C.; Etoh T.; Landry B. & Matsuzoe N. (1997). RAPD markers related to pollen fertility in garlic (*Allium sativum* L.) *Breeding science*, Vol. 74, pp. 359-362

Hou, A. & Peffley E.B. (2000). Recombinant chromosomes of advanced backcross plants between *Allium cepa* L. and *A. fistulosum* L. revealed by *in situ* hybridization. *Theor Appl Genet*, Vol. 100, pp. 1190-1196

Jones, G.H.; Khazanehdhari K.A & Ford-Lloyd B.V. (1996). Meiosis in leek (*Allium porrum* L.) revisited. II. Metaphase I observations. *Heredity*, Vol. 76, pp. 186-191

Jones, H.A. & Clarke A.E. (1942). A natural amphidiploid from an onion species hybrid *Allium cepa* x *Allium fistulosum* L. *Heredity*, Vol. 33, pp. 25-32

Jones, H.A. & Mann L.K. (1963). Onions and Their Allies - Botany, Cultivation, and Utilization. Interscience Publishers Inc., ISBN 100249388731, New York3.

Jones, R.N. & Rees H. (1968) Nuclear DNA variation in *allium*. *Heredity*, Vol. 23, pp. 591-605

Jones, R.N. (1983). Cytogenetic evolution in the genus *Allium*. In: *Cytogenetic of crop plants*. Swaminathan M.S.; Gupta P.K. & Sinha V., pp. 516-554, Macmillan Press Ltd, ISBN 0333904230, India

Jones, R.N. (1990). Cytogenics. In: *Onions and Allied crops. Vol I. Botany, physiology and genetics*, Rabinowitch H.D. & Brewster J.L., pp. 199-214, CRC Press, ISBN 0849363004, Florida

Kalkman, E.R. (1984). Analysis of the C-banded karyotype of *Allium cepa* L. Standard system of nomenclature and polymorphism. *Genetica*, Vol. 65, pp. 141-148

Keller, R.J.; Schubert I.; Fuchs J. & Meister A. (1996). Interspecific crosses of onion with distant *Allium* species and characterization of the presumed hybrids by means of flow cytometry, karyotype analysis and genomic *in situ* hybridization. *Theor Appl Genet*, Vol. 92, pp. 417-424

Keusgen, M. (2002). Health and *Alliums*. In: *Allium Crop Science: recent advances*. Rabinowitch H.D. & Currah L., pp. 537-379, CABI Publishing, ISBN 0851995101, UK

Khazanedhari, K.A. & Jones G.H. (1997). The causes and consequences of meiotic irregularity in the leek (*Allium ampeloprasum* spp. *Porrum*); implications for fertility, quality and uniformity. *Euphytica*, Vol. 93, pp. 313-319

Khazanehdhari, K.A.; Jones G.H. & Ford-Lloyd B.V. (1995). Meiosis in the leek (*Allium porrum* L.) revisited. I. Prophase I pairing. *Chromosome Research*, Vol. 3, pp. 433-439

Khrustaleva L.I. & Kik C. (2000) Introgression of *Allium fistulosum* into *A. cepa* mediated by *A. roylei*. *Theor Appl Genet*, Vol. 100, pp. 17–26

Khrustaleva, L.I. & Kik C. (1998) Cytogenetical studies in the bridge cross *Allium cepa* x (*A. fistulosum* x *A. roylei*). *Theor Appl Genet*, Vol. 96, pp. 8–14

Khush, G.S. & Brar D.S. (1992). Overcoming the barriers in hybridization. In: *Distant hybridization of crop plants*, Monographs on Theoretical and Applied Genetics. Vol 16., Kaloo G. & Chowdhury J.B., pp. 47-61, Springer Verlag, ISBN 3540531734, Berlin

Kiełkowska, A. & Adamus A. (2006). Growth of pollen tubes from foreign species in carrot (*Daucus carota* L.) pistils. In: *Haploids and doubled haploids in genetics and plant breeding.* Adamski T. & Surma M., pp. 193-197, Prodruk, ISBN 8389887444, Poland

Kiełkowska, A. & Adamus A. (2010). Morphological, cytological and molecular evaluation of interspecific F1 (*A.galanthum* x *A.cepa*) hybrids. *Biotechnologia,* Vol. 2(89), pp. 146-155

Kik, C. (2002). Exploitation of wild relatives for the breeding of cultivated *Allium* species. In: Allium Crop Science: recent advances. Rabinowitch H.D. & Currah L., pp. 81-100, CABI Publishing, ISBN 0851995101, UK

Kirti, P.B., Prakash S. & Chopra V.L. (1991). Interspecific hybridization between *Brassica juncea* and *B. spinescens* through protoplast fusion. *Plant Cell Rep,* Vol. 9, pp. 639-642

Klima, M.; Vyvadilov☐ M. & Kučera V. (2008). Chromosome doubling effects of selected antimitotic agents in *Brassica napus* microspore culture. *Czech J Genet Breed,* Vol. 44, pp. 30-36

Konvička, O. & Levan A. (1972). Chromosome studies in *Allium sativum.* Hereditas, Vol. 72, pp. 129-148

Koul, A.K. & Gohil R.N. (1970). Cytology of the tetraploid *Allium ampeloprasum* with chiasma localization. *Chromosoma,* Vol. 29, pp. 12-19

Labani, R. & Elkington T. (1987). Nuclear DNA variation in the genus *Allium* L. (*Liliaceae*). *Heredity,* Vol. 59, pp. 119-128

Levan, A. (1931). Cytological studies in Allium. A preliminary note. Hereditas, Vol.15, pp. 347-356

Levan, A. (1932). Cytological studies in *Allium* II. Chromosome morphological contribution. *Hereditas,* Vol. 16, pp. 57-299

Levan, A. (1933). Cytological studies in *allium* IV. *Allium fistulosum. Svensk Bot Tids,* Vol. 27, pp. 211-232

Levan, A. (1935). Cytological studies in *Allium* V. The chromosome morphology of some diploid species of *Allium. Hereditas,* Vol. 20, pp. 289-330

Levan, A. (1936) Die zytologie von *Allium cepa* x *fistulosum. Hereditas,* Vol. 21, pp. 195-214

Levan, A. (1940). Meiosis of *Allium porrum,* a tetraploid species with chiasma localization. *Hereditas,* Vol. 26, pp. 454-462

Levan, A. (1941). The cytology of the species hybrid *Allium cepa* x *fistulosum* and its polyploidy derivatives. *Hereditas,* Vol. 27, pp. 253-272

Maeda, T. (1937). Chiasma studies in *Allium fistulosum, Allium cepa* and their F1, F2 and backcross hybrids. *Jap J Genet,* Vol. 13, pp. 146-159

Manickam, S. & Sarkar K.R. (1999a). Foreign pollen tube growth in maize after chemical treatments. *Indian J Gen & Plant Breed,* Vol. 59(1), pp. 53-58

Manickam, S. & Sarkar K.R. (1999b). Maize, pearl millet and sorghum pollen tube growth rate in maize silk. *Ann Agricult Research,* Vol. 20(2), pp. 216-219

McCollum, G.D. (1971). Sterility of some interspecific *Allium* hybrids. *J Am Soc Hort Sci,* Vol. 96, pp. 359-362

McCollum, G.D. (1980). Development of the amphidiploids of *A. galanthum* x *A. cepa,* J *Heredity,* Vol. 71, pp. 445-447

McCoy, T.J. & Echt C.S. (1993). Potential of trispecies bridge crosses and random amplified DNA markers for introgression of *Medicago daghestanica* and *M. pironae* germplasm in to alfalfa (*M. sativa*). *Genome*, Vol. 36, pp. 594-601

Murin, A. (1964). Chromosome study in *Allium porrum* L. *Caryologia* Vol, 17, pp. 575-578

Nomura, Y. & Makara K. (1993). Production of interspecific hybrids between Rakkyo (*Allium chinense*) and some other *Allium* species by embryo rescue. *Jpn J Breed*, Vol. 43, pp. 13–21

Nomura, Y. & Oosawa K. (1990). Production of interspecific hybrids between *Allium chinense* and *Allium thunbergii* by in ovulo embryo culture. *Jpn J Breed*, Vol. 40, pp. 531-535

Nomura, Y.; Kazuma T.; Makara K. & Nagai T. (2002). Interspecific hybridization of autumn-flowering *Allium* species with ornamental *Alliums* and characteristics of the hybrid plants. *Sci Hort*, Vol. 95, pp. 223-237

Nomura, Y.; Maeda M.; Tsuchiya T. & Makara K. (1994). Efficient production of interspecific hybrids between *Allium chinense* and edible *Allium* spp. through ovary culture and pollen storage. *Breed Sci*, Vol. 44, pp. 151-155

Ohsumi, C.A.; Kojima K.; Hinata K.; Etoh T. & Hayashi T. (1993). Interspecific hybrid between *Allium cepa* and *Allium sativum*. *Theor Appl Genet*, Vol. 85, pp. 969-975

Orton, T.J. & Steidl P.J. (1980). Cytogenetic analysis of plants regenerated from colchicine-treated callus cultures of an interspecific *Hordeum* hybrid. *Theor Appl Genet*, Vol. 57, pp. 89-95

Peffley, E.B. & Mangum P.D. (1990) Introgression of *Allium fistulosum* L. into *Allium cepa* L: cytogenetic evidence. *Theor Appl Genet*, Vol. 79, pp. 113-118

Peffley, E.B. (1986). Evidence for chromosomal differentiation of *A. fistulosum* and *A. cepa*. *J Am Soc Hort Sci*, Vol. 111, pp. 126-129

Peffley, E.B.; Corgan J.N.; Horak H.G. & Tanksley S.D. (1985). Electrophoretic analysis of *Allium* alien addition lines. *Theor Appl Genet*, Vol. 71, pp. 176-184

Peterka, H.; Budahn H. & Schrader O. (1997). Interspecific hybrids between onion (Allium cepa L.) with S-cytoplasm and leek (*Allium ampeloprasum* L.). *Theor Appl Genet*, Vol. 94, pp. 383-389

Peterka, H.; Budahn H.; Schrader O. & Havey M.J. (2002). Transfer of male-sterility-inducing cytoplasm from onion to leek (*Allium ampeloprasum*). *Theor Appl Genet*, Vol. 105, pp. 173-181

Peters, R.J.; Netzer D. & Rabinowitch H.D. (1984). A progress report: pink root resistance in *Allium cepa* x *Allium fistulosum* L hybrids and progeny. In: *Proc. 3rd Allium Eucarpia Symp.*, Wageningen, pp. 70-73, The Netherlands

Pich, U.; Fritsch R. & Schubert I. (1996) Closely related *Allium* species (*Alliaceae*) share a very similar satellite sequence. *Plant Systematics and Evol*, Vol. 202, pp. 255-264

Rabinowitch, H.D. (1997) Breeding alliaceous crops for pest resistance. *Acta Hort*, Vol. 433, pp. 223-246

Sastri, D.C. & Schivanna K.R. (1976). Attempts to overcome interspecific incompatibility in *Sesanum* by using of recognition pollen. *Ann Bot*, Vol. 40, pp. 891-893

Schrader, O.; Budahn H.; Ahne R. & Peterka H. (2000). Cytogenetic and molecular analysis of somaclonal variants in *Allium cepa* x *A. ampeloprasum* hybrid. *Vortr Pflanzenzuecht*, Vol. 47, pp. 53, ISSN 0723-7812

Schwarzacher, T.; Leith A.R.; Benett M.D. & Heslop-Harrison J.S. (1989). *In situ* localization of parental genomes in a wide hybrid. *Ann Bot* Vol. 64, pp. 315-324

Settler, R.F. (1968). Irradiated mentor pollen: Its use in remote hybridization of the cottonwood. *Nature*, Vol. 219, pp. 746-747

Shigyo, M.; Imamura K. Iino M.; Yamashita K., & Tashiro Y. (1998). Identification of alien chromosomes in a series of *Allium fistulosum* – *A. cepa* monosomic addition lines by means of genomic in situ hybridization. *Genes Genet Syst*, Vol. 73, pp. 311-315

Shimonaka, M.; Hosoki T.; Tomita M. & Yasumuro Y. (2002). Production of somatic hybryd plants between Japanese bunching onion (*A. fistulosum* L.) and bulb onion (*A. cepa* L.) *via* electrofusion. *J Jpn Soc Hort Sci*, Vol. 71, pp. 623-631

Smith, H.H.; Kao K.N. & Combatti N.C. (1976). Interspecific hybridization by protoplast fusion in *Nicotiana*. *The J Heredity*, Vol. 67, pp. 123-128

Song, P.; Kang W. & Peffley E.B. (1997). Chromosome doubling of *Allium fistulosum x A. cepa* interspecific F1 hybrids through colchicine treatment of regenerating callus. *Euphytica*, Vol. 93, pp. 257-262

Stack, S.M. & Roelofs D. (1996) Localized chiasmata and recombination nodules in the tetraploid onion *Allium porrum*. *Genome*, Vol. 39, pp. 770-783

Stevenson, M.; Armstrong S.J.; Ford-Lloyd B.V. & Jones G.H. (1998). Comparative analysis of crossover exchanges and chiasmata in *Allium cepa x fistulosum* after genomic in situ hybridization (GISH). *Chromosome Research*, Vol. 6, pp. 567-574

Sukno, S.; Ruso J.; Jan C.C; Melero-Vara J.M. & Fernandez-Martinez J.M. (1999). Interspecific hybridization between sunflower and wild perennial *Helianthus* species *via* embryo rescue. *Euphytica*, Vol. 106, pp. 69–78

Sybenga, J. (1996). Recombination and chiasmata: few but intriguing discrepancies. *Genome*, Vol. 39, pp. 473-484

Taylor, R.W. (1925). The chromosome morphology of *Velthemia*, *Allium* and *Cyrtanthus*. *Am J Bot*, Vol. 12, pp. 104-115

Thomas, H.M.; Morgan W.G.; Meredith W.G.; Humphreys M.W.; Thomas H. & Legget JM (1994). Identification of parental and recombined chromosomes in hybrid derivatives of *Lolium multiflorum* x *Festuca pratensis* by genomic *in situ* hybridization. *Theor Appl Genet*, Vol. 88, pp. 903-913

Toole, M.G. & Clarke A.E. (1994). Chromosome behavior and fertility of colchicine-induced tetraploids in *A. cepa* and *A. fistulosum*. *Herbertia*, Vol. 11, pp. 295-303

Ulloa, M.; Corgan J.N. & Dunford M. (1994). Chromosome characteristics and behavior differences in *Allium fistulosum* L., *A. cepa* L., their F1 hybrid, and selected backcross progeny. *Theor Appl Genet*, Vol. 89, pp. 567-571

Ulloa, M.; Corgan J.N. & Dunford M. (1995). Evidence for nuclear-cytoplasmic incompatibility between *Allium fistulosum* and *A. cepa*. *Theor Appl Genet*, Vol. 90, pp. 746-754

Umehara, M.; Sueyoshi T. & Shimomura K. (2006). Interspecific hybrids between *Allium fistulosum* and *Allium schoenoprasum* reveal carotene-rich phenotype. *Euphytica*, Vol. 148, pp. 295–301

Umehara, M.; Sueyoshi T.; Shimomura K.; Hirashima K.; Shimoda M. & Nakahara T. (2007) Production and characterization of interspecific hybrids between *Allium fistulosum* L. and *Allium schoenoprasum* L. *Bull of the Fukuoka Agric Res Center*, Vol. 26, pp. 25-30

van der Meer, Q.P. & van Bennekom J.L. (1978). Improving the onion crop (*Allium cepa* L.) by transfer of characters from *A. fistulosum*. *Biuletyn Warzywniczy*, Vol. 22, pp. 87-91 van der Valk P.; de Vries S.E.; Everink J.T.; Verstappen F. & de Vries J.N. (1991a). Pre- and post-fertilization barriers to backcrossing the interspecific hybrid between *Allium fistulosum* L. and *A. cepa* L. with *A. cepa. Euphytica,* Vol. 53, pp. 201-209

van der Valk, P.; Kik C.; Verstappen F.; Evernik J.T. & de Vries J.N. (1991b). Independent segregation of two isozyme markers and inter-plant differences in nuclear DNA content in the intespecific cross (*Allium fistulosum* L. x *A. cepa* L.) x *A. cepa* L. *Euphytica,* Vol. 55, pp. 151-156

van Heusden, A.W.; van Ooijen J.W.; Vrielink-van Ginkel M.; Verbeek W.H.J.; Wietsma W.A. & Kik C. (2000). A genetic map of an interspecific cross in *allium* based on amplified fragment length polymorphism (AFLP) markers. *Theor Appl Genet,* Vol. 100, pp. 118-126

Ved Brat, S. (1965a). Genetic systems in *Allium* I. Chromosome variation. *Chromosoma,* Vol. 16, pp. 486-49

Ved Brat, S. (1965b). Genetic systems in *Allium* III. Meiosis and breeding systems. *Heredity,* Vol. 20, pp. 325-339

Ved Brat, S. (1967). Genetic systems in *Allium* IV. Balance in hybrids. *Heredity,* Vol. 22, pp. 387-396

Wan, Y.; Petolino J.F. & Widholm J.M. (1989). Efficient production of doubled haploid plants through colchicine treatment of anther derived maize callus. *Theor Appl Genet,* Vol. 77, pp. 889-892

Williams, E.G.; Verry I.M. & Williams W. M. (1982). Use of embryo culture in interspecific hybridization. In: *Plant Improvement and Somatic Cell Genetics.* Vasil I.K.; Scowcroft W.R. & Frey K.J., pp. 119–128, Academic Press, ISBN 0127149805, New York

Yamashita, K.; Hisatsune Y.; Sakamoto T.; Ishizuka K. & Tashiro Y. (2002). Chromosome and cytoplasm analyses of somatic hybrids between onion (*Allium cepa* L.) and garlic (*A. sativum* L.). *Euphytica,* Vol. 125, pp. 163-167

Yanagino, T.; Sugawara E. & Watanabe M. (2003). Production and characterization of an interspecific hybrid between leek and garlic. *Theor Appl Genet* Vol. 107, pp. 1–5

Zenkteller, M. (1990) *In vitro* fertilization and wide hybridization in higher plants. *Plant Sci,* Vol. 9, pp. 267-279

Zenkteller, M.; Bagniewszka-Zadworna A. & Zenkteller E. (2005). Embryological studies on ovules of *Melandrium album* pollinated *in vitro* with *Lychnis coronaria* pollen grains. *Acta Biol Cracoviensia series Bot,* Vol. 47, pp. 135-138

Zhao, J. & Simmonds D.H. (1995). Application of trifluralin to embryogenic microspore cultures to generate doubled haploid plants in *Brassica napus. Physiol Plantarum,* Vol. 95, pp. 304–309

4

Cohesins and Cohesin-Regulators in Meiosis

Adela Calvente and José L. Barbero

Cell Proliferation and Development Department, Centro de Investigaciones Biológicas,
Spain

1. Introduction

Cells have rigorous mechanisms controlling sister chromatid cohesion during cell division to ensure proper distribution of genetic material to daughter cells. Errors in these mechanisms often lead to aneuploidy, frequently implicated in cell death, tumour development and infertility. The main actor in this process is a four-protein complex called the cohesin complex. The role of cohesin complex in chromosome segregation is mediated by the formation of a ring-like structure, which entrapped replicated DNA. The dynamic of the cohesin ring is regulated by a still undetermined number of cohesin-interacting proteins. These cohesin-regulators were essentially identified and studied in relation with the cohesion function of cohesin complexes. In the last years, we have improved our understanding of the key players in the regulation of sister chromatid cohesion during cell division in mitosis and meiosis. During meiosis the formation and disassemble of synaptonemal complex (SC), the recombination between homologue chromosomes and the maintenance of sister chromatid cohesion until metaphase II - anaphase II transition require unique features and players participating in the control of chromosome segregation. While in mammalian mitosis there are essentially two cohesin complexes, which only differ in the STAG subunit (STAG1 or STAG2), in meiosis several cohesin complexes composed by specific and unspecific meiotic cohesins coexist and probably they develop different functions depending of their spatio-temporal chromosome localization. On the other hand, a correct control of chromosome cohesion in meiosis involves specific regulators that monitor the sequential cohesin release during both meiotic divisions. Recently excellent papers have appeared in the literature looking in depth on the molecular control of sister chromatid cohesion and cohesins in cell division. In this chapter, we review the implication of cohesins and cohesin-interacting proteins in meiosis-specific processes and chromosome dynamic. Furthermore due to the increasing relevance of cohesins in human syndromes, we briefly point how problems in their tasks during mammalian mitotic and meiotic cycle drive to pathological situations.

2. Cohesin basic concepts

The pair of sister chromatids produced after DNA replication must be maintained together throughout the G2 phase and until its segregation to ensure a correct cell division. Thus, the sister chromatid cohesion is established during S phase (Miyazaki & Orr-Weaver, 1994) and although in mitosis the cohesion is released during prophase and prometaphase in arms (Hauf et al., 2005), the sister chromatid are joined at centromeres until the onset of anaphase. In meiosis, there are two consecutive chromosome divisions, segregating homologous chromosomes in anaphase I and sister chromatids in anaphase II. Thus the cohesion regulation presents specific characteristics of this kind of cell division, and whereas the arm cohesion is lost at anaphase I, the centromere cohesion is released at anaphase II (Watanabe, 2004). Since both cohesion releases follow the same pathway in meiosis, the meiotic centromere cohesin complexes must be protected until the second meiotic division (Kitajima et al., 2004). At any case, mitosis and meiosis, the multi-protein complex responsible of sister chromatid cohesion is called cohesin complex and it was first characterized in *Saccharomyces cerevisiae* and *Xenopus laevis* (Losada et al., 1998, Michaelis et al., 1997). The cohesin complex is composed by four subunits: two structural maintenance of chromosomes family proteins (SMC1 and SMC3), one α-kleisin subunit (SCC1/RAD21), and a HEAT-repeat domain protein (SCC3/SA/STAG) (Nasmyth & Haering, 2005). The most of components of the cohesin complex are conserved from yeast to humans (Uhlmann, 2001) but there are specific subunits depending on the species or the cell division type. The most conserved cohesins are the SMC proteins, which form a V-shaped heterodimer, representing the core of the cohesin complex (Haering et al., 2002). Higher eukaryotes have two mitotic SA/STAG family members, SA1/STAG1 and SA2/STAG2 (Carramolino et al., 1997), which do not coexist and are present in different cohesin complexes (Losada et al., 2000). In germ cells have been characterized distinct meiosis-specific subunits of cohesin complex in different organisms. In mammals, a meiotic paralogue of SMC1 has been described, the SMC1β, thus the subunit presents in mitosis and meiosis is called SMC1α (Revenkova et al., 2001). In yeast and mammals REC8 is the meiotic paralogue of SCC1/RAD21 subunit (Eijpe et al., 2003, Molnar et al., 1995, Watanabe & Nurse, 1999) and in mice a new α-kleisin has been identified recently, the RAD21L, a paralogue of RAD21 (Gutierrez-Caballero et al., 2011, Ishiguro et al., 2011, Lee & Hirano, 2011). RAD21L interacts with STAG3 (meiosis-specific SCC3 subunit) and with the three described SMC cohesin subunits, SMC1α, SMC1β and SMC3. STAG3 is the meiosis-specific paralogue for STAG1/2 in mammals (Pezzi et al., 2000, Prieto et al., 2001). Whereas these cohesin subunits, RAD21L, REC8, STAG3 and SMC1β, are meiosis-specific and they are present specifically in spermatocytes and oocytes (Garcia-Cruz et al., 2010, Herran et al., 2011, Prieto, et al., 2001), different cytological and molecular analysis show the participation of the SMC1α, RAD21 and STAG2 in mammalian mitosis and meiosis (Prieto et al., 2002, Revenkova et al., 2004, Revenkova & Jessberger, 2006). Characterization of distinct meiotic cohesin complexes containing REC8, RAD21 or RAD21L as α-kleisin subunits, which have distinct localization patterns and dynamics, as well as the simultaneous presence of SMC1α and β-containing complexes and the presence of STAG2 and STAG3 in mammalian meiosis suggest a large variety of putative cohesin complexes formed by combinations of cohesin subunits (Suja & Barbero, 2009). The most accepted model of the cohesin complex organization describes a heterodimer of SMC proteins jointed by their hinge domains. The ATPase heads of SMC1 and SMC3 are connected by the α-kleisin, forming a tripartite ring and finally the SCC3 subunit interacts with SCC1 via SCC1's C-terminus (Fig. 1A) (Haering, et al., 2002). Two major models, which are not mutually exclusive, have been proposed for the cohesin complex function in sister chromatid cohesion and its interaction with the DNA molecules. The first one is based on the electronic microscopy results and structural characteristics of SMCs and described a cohesin complex forming a ring-like structure (Fig. 1B), which mediates cohesion by embracing chromatin fibers of both sister chromatids (Gruber et al., 2003). The second model was named handcuff model, it involves the participation of two cohesin rings formed by SMC1/SMC3/SCC1 subunits which interacts in an SCC3-dependent manner (Fig. 1C). In this model each sister chromatid is encircled by a tripartite cohesin ring (Zhang et al., 2008a).

The activity of the cohesin complex is closely related to the action of three cohesin cofactors: PDS5, WAPL (Fig. 1A) and Sororin. PDS5 is associated with cohesins but in a less tightly bound manner than the cohesin complex proteins. In vertebrates there are two PDS5 homologues, PDS5A and PDS5B. The role of PDS5 is related to the maintenance of cohesion and the modulation of the interaction of cohesin complex with the chromatin (Losada et al., 2005). WAPL is involved in heterochromatin organization in *Drosophila melanogaster* (Verni et al., 2000) and in human cells regulates the resolution of sister chromatid cohesion during prophase (Gandhi et al., 2006).

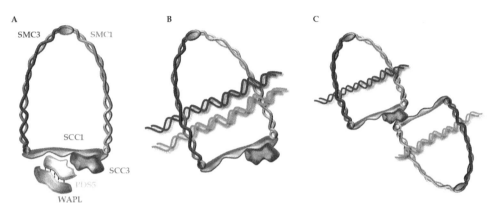

Fig. 1. Cohesin complex organization and cohesion models.
A. The cohesin complex is formed by four proteins: SMC1, SMC3, SCC1 and SCC3 (green, red, orange and blue piece respectively). The SMC1, SMC3 and SCC1 subunits form a ring-like structure. The SCC3 protein interacts with SCC1 to complete the cohesin complex. The cofactors PDS5 (yellow piece) and WAPL (purple piece) interact to form a protein complex, which is associated to the cohesin complex by SCC3 interaction. **B.** Ring-embraced model of cohesion. A single cohesin complex hugs the two sister chromatids. **C.** Handcuff model of cohesion. Two cohesin rings interacts by a single SCC3 molecule. Each cohesin ring encircles a single chromatid.

Regarding to the Sororin, it has been described in culture cells that this protein have a role in centromere cohesion, probably it is implicated in the protection of centromeric cohesin complexes (Diaz-Martinez et al., 2007). Moreover Sororin seems to have a role in the cohesion establishment during replication (Lafont et al., 2010). The implication of all these cofactors in meiosis is a wide field of study nowadays. The importance of cohesins, cohesin complex and cohesion cofactors not only consists in keeping jointed both sister chromatid and then ensure the bipolar attachment and the chromosomal segregation (Tanaka et al., 2000) in mitosis and meiosis, but the cohesins acquire a pivotal role in many different aspects of meiosis and chromosome dynamics such as the modulation of gene expression, the double strand breaks (DSBs) repair, axial elements (AEs) formation or pairing and synapsis of homologous chromosomes. In this chapter some of those different roles of cohesins and cohesion cofactors are reviewed, giving special attention to the specificity of some cohesin subunits to specific functions.

3. Loading, establishment, maintenance and release of cohesin complexes

Chromosome segregation control is managed by an intricate network of events, which assures that each daughter cell receives the right chromosome number, equal amount of chromosomes in mitosis and a haploid recombined chromosomal pack in meiosis. On one hand, fails in mitotic chromosome segregation compromises the viability of the organism, on the other hand fails in meiotic chromosome segregation jeopardizes the fertility of individuals. The importance of cohesion must be taken into account from the loading of cohesins to the establishment of cohesion, maintenance of cohesive function and finally the removal of cohesin complexes. In this context, the association/dissociation of sister chromatids is a well-controlled process that involves the role of different cohesin-interacting proteins. The loading of cohesin complexes to chromosomes occur near the G1 to S phase previous to cell division, thus it would be easy to think that the regulators of cohesin loading would be similar in mitosis and meiosis. In fact, most of them have been first described in yeast mitosis and afterwards a similar meiotic function was identified in some cases. Although we do not should forget that the transition from mitosis to meiosis could be trigger from the premeiotic S phase, and taking into account the presence of meiotic cohesin complexes and the special regulation of meiotic cohesion release, crucial differences might exist between mitosis and meiosis. It was described in *S. cerevisiae* that the loading of cohesin complexes depends on the SCC2/SCC4 adherin complex (Ciosk et al., 2000). This loading is not sufficient for the cohesion function in chromosome segregation and the Eco1/Ctf7p acetyltransferase is required for the establishment of cohesion in mitotic yeast chromosomes (Skibbens et al., 1999, Toth et al., 1999). In humans, two Eco1 orthologues have been identified: ESCO1 and ESCO2. Both proteins exhibit not redundant acetyltransferase activity and the depletion of any of them by siRNA cause defects in sister chromatid cohesion and chromosome segregation. However, in S-phase the role of both proteins seems to be directly related to the cohesion establishment and do not with the cohesin loading (Hou & Zou, 2005). The substrate of Eco1/ESCO1 is SMC3 in both human and yeast cells, being required the acetylation of SMC3 for sister chromatid cohesion. The acetylation of SMC3 could regulate its interaction with SCC1, promoting the turn from a chromosome-bound cohesin complex into a cohesive complex (Zhang et al., 2008b), suggesting that the cohesin complexes with acetylated SMC3 may be stably bound to chromosomes (Unal et al., 2008). The activity of Eco1 could be facilitated due to its interaction with PCNA (proliferating cell nuclear antigen) (Moldovan et al., 2006). The acetylation of SMC3 is maintained until metaphase - anaphase transition and its deacetylation in yeast mitosis depends on Hos1, but this deacetylase cannot act until the SCC1 cohesin subunit is cleaved by a protease called Separase, recycling then the tripartite open ring of cohesin (Beckouet et al., 2010, Borges et al., 2010). In mammalian mitosis the deacetylase responsible for SMC3 deacetylation has not been identified, moreover the cycle of SMC3 acetylation/deacetylation in meiosis is still poorly known.

The release of cohesion in mitotic prophase and anaphase follows different pathways of resolution (Waizenegger et al., 2000). In mitosis, the arm cohesion is released during prophase in a Separase-independent manner, it depends on the phosphorylation of SCC3/SA2/STAG2 subunit by Aurora B and Polo-like kinases (Hauf, et al., 2005), but the correct chromosome segregation depends on the centromeric cohesion, which is maintained until anaphase onset and then released by the Separase activity. The release of cohesin complexes from centromeric chromatin at mitotic metaphase - anaphase transition is mediated by the Separase, a specific protease that cleaves the SCC1 subunit of the cohesin complex (Uhlmann et al., 1999), destabilizing the association of cohesins to chromatin. Before anaphase, Separase remains inactivated by binding to its specific inhibitor Securin (Ciosk et al., 1998). Activation of the anaphase promoting complex/cyclosome (APC/C) leads to ubiquitination of Securin, allowing cleavage of SCC1/RAD21 by Separase and triggering the onset of anaphase (Uhlmann et al., 2000). In meiosis there are two consecutive chromosome segregations which are triggered by a Separase-dependent mechanism of lost of cohesion. During first meiotic division, the cohesin complexes at chromosome arms are removed during the metaphase I - anaphase I transition, allowing segregation of recombined homologues to opposite poles. In the second meiotic division, the cohesion is released from centromeres and sister chromatids are segregated to opposites poles to generate haploid gametes. Then, this type of cell division presents unique characteristics in sister chromatid cohesion removal and centromere cohesion protection. This is necessary to prevent premature separation of sister chromatids in order to avoid aneuploidy in the resulting gametes. Related to the centromere protection a protein family was indentified in fission yeast, the Shugoshins: Sgo1 and Sgo2 (Kitajima, et al., 2004). In these organisms Sgo1 seems to be essential in meiosis and Sgo2 is mostly implicated in mitotic division. The activity of Sgo1 and Sgo2 is related to the recruitment of a serine/threonine protein phosphatase 2A (PP2A) (Kitajima et al., 2006). In humans these proteins are called SGOL1 and SGOL2 and in mitosis they also collaborate with PP2A. In mice it has been described the localization of SGOL2 in male meiosis and it corresponds with a protection function of centromeric cohesion during first meiotic division (Gomez et al., 2007). An interesting model is the SGOL2-deficient mouse, where a precocious dissociation of the meiosis-specific REC8 cohesin complexes from anaphase I centromeres was observed (Llano et al., 2008), demonstrating the specific implication of SGOL2 in centromere protection during meiosis.

The cohesin-regulators PDS5 and WAPL are also involved in the opening/closing of cohesin ring by interactions with different cohesin subunits. These cofactors are not required for cohesin association to chromosomes but they are necessary for cohesin complex dynamics. The action of PDS5 is related to its interaction with the SA1 and SA-containing complexes in somatic cells (Sumara et al., 2000). In vertebrates two PDS5 proteins have been characterized: PDS5A and PDS5B. Both are large HEAT-repeat proteins that bind to chromatin in a cohesin-dependent manner in human cells and *Xenopus* egg extracts. RNAi depletion of PDS5A and PDS5B show that both are needed for maintaining cohesion, altering preferentially centromeric cohesion in *Xenopus* egg extracts. (Losada, et al., 2005). Mice lacking PDS5B function die shortly after birth and exhibit multiple developmental anomalies that resemble those found in humans with Cornelia de Lange syndrome (CdLS), indicating a relevant function for PDS5B beyond chromosome segregation, but there are no discernible defects in sister chromatid cohesion (Zhang et al., 2007). Despite of these contradictory results, the function of PDS5 has been related with the maintenance of sister chromatid cohesion during G2 (Panizza et al., 2000). Although, whether PDS5 also has a function in sister chromatid resolution remains to be determined. Another interesting cohesin cofactor is the product of the previously identified *Drosophila* wings apart-like (*Wapl*) gene, involved in heterochromatin organization (Verni, et al., 2000). Human WAPL regulates the resolution of sister chromatid cohesion and promotes cohesin complex removal by direct interaction with the RAD21 and SA/STAG cohesin subunits (Gandhi, et al., 2006). Thus, WAPL seems to destabilize cohesins. It has been proposed that PDS5 and WAPL form a protein complex, which in association with SCC3 cohesin subunit antagonizes the establishment of cohesion, calling the WAPL-PDS5 anti-establishment complex.

Skibbens proposed a model of cohesion establishment where the recruitment of PDS5-WAPL complex onto SCC3 subunit prevents the binding of cohesin complexes by destabilizing cohesin-cohesin association. During S phase, Eco1/ESCO1 acetylates SMC3, which inhibits the WAPL-PDS5 activity temporally (Skibbens, 2009). After S phase, PDS5 would stabilize the cohesin complexes activity.

Sororin has been implicated in centromere cohesion. This protein was firstly identified in a screen for substrates of the APC in vertebrates and no homologues have been described in other organisms. Different results in somatic cells suggested that Sororin interacts with the cohesin complex and it is essential for the maintenance of sister chromatid cohesion. Sororin is ubiquitinized and degraded after sister chromatid cohesion is dissolved (Rankin, 2005). Studies on Sororin-depleted and Shugoshin-depleted cells indicate that both proteins might act in concert in the protection of centromeric cohesion (Diaz-Martinez, et al., 2007). Sororin is also needed for maintaining stable chromatin-bound cohesin and DSBs repair in G2 (Schmitz et al., 2007). The Sororin recruitment depends on Eco2/ESCO2, both are subtrates of APC and its activity is related to the DNA replication (Lafont, et al., 2010). In agreement with these findings, Nishiyama reported that DNA replication and cohesin acetylation promote binding of Sororin to cohesin complex and that Sororin displaces WAPL from its binding partner PDS5, thus it would contribute to maintenance of a stable binding of cohesin to chromatin (Nishiyama et al., 2010).

Despite of the relevant roles in chromosome cohesion control suggested by all these results, there are few data regarding to the putative functions of cohesion cofactors such as SCC2/SCC4, Eco1/ESCO1, Eco2/ESCO2, WAPL, PDS5 or Sororin in chromosome segregation in meiosis.

4. Roles of the cohesin-regulators in meiosis

Updates there are no many evidences regarding to the specific implication of all the cohesion cofactors in meiosis. It is predictable that all of them might act in meiosis in a similar way to that described in mitotic cells or perhaps they acquire specific roles due to the particular regulation of meiotic division. Update the unique indication that ESCO2 have a role in meiosis was the identification of *Esco2* gene as a candidate to be a potential regulator of the transition from mitosis to meiosis in mammals, identifying by means microarray database in both testis and ovary of mouse (Hogarth et al., 2011). The data are supported not only by a pattern of mRNA expression but also by protein immunolocalization, showing on one hand that *Esco2* is expressed in testis and ovary, specifically in the embryonic gonad, and on the other hand that ESCO2 localized to the nucleus of spermatocytes at early prophase I. This last probe was performed by immunofluorescence techniques over testis sections at different ages. At 10dpp (days post-partum) male mice, ESCO2 was presented at preleptotene and leptotene diffusely and from 15dpp this protein is located at pachytene in a discrete domain within the nuclei. Our laboratory is actually studying the ESCO2 implications in meiosis and using the same antibody (Bethyl Laboratories, A301-689A) we have identified that the discrete domain observed by Hogarth et al., is in fact the XY body (Figs. 2 and 3). We have analyzed the male mice meiosis over squashes of seminiferous tubules (Page et al., 1998, Parra et al., 2004) and spreads of meiocytes (Peters et al., 1997, Viera et al., 2009a) and performed the double-immunolocation of ESCO2 and SYCP3 (Santa Cruz Biotechnology, sc-74569). Our results indicate that the ESCO2 acetyltranferase is present over the chromatin of XY body from zygotene to late stages of prophase I, detecting a more intense signal at pachytene (Fig. 2 and 3 A). We distinguish the XY body in squashed spermatocytes because of it usually is located at nuclear periphery and its AEs are partially unsynapsed due to this pair of chromosomes share only a homologue region called pseudoatosomal region, which the SC is exclusively formed in. Also the DAPI chromatin staining led us to identify the XY body undoubtedly in squashes (Fig. 2D). At pachytene nucleus the synapsis between autosomes is completed, which can be detected by SYCP3 immunolabeling (Fig. 2 and 3 B).

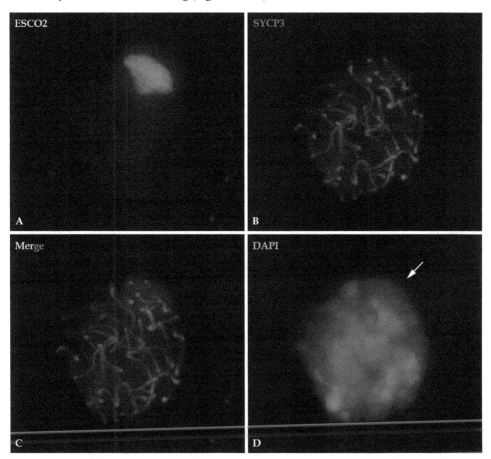

Fig. 2. Immunolabeling of ESCO2 (green) and SYCP3 (red) at pachytene after squasing of spermatocytes.
A. ESCO2 immunolabeling (rabbit anti-ESCO2 Bethyl Laboratories, A301-689A; 1:50 dilution) mark the XY body at pachytene. FITC-conjugated anti-rabbit IgG secondary antibody (Jackson Laboratories) was used at 1:150 dilution. **B.** SYCP3 (mouse monoclonal antibody anti-SYCP3. Santa Cruz Biotechnology, SCP-3 (D-1): sc-74569; 1:100 dilution) mark the LEs of synapsed autosomes and the AE/LEs of the XY body. DyLight594-conjugated anti-mouse IgG secondary antibody (Jackson Laboratories) was used at 1:150 dilution. **C.** The merge image of ESCO2 and SYCP3 signals is shown. **D.** The chromatin was

counterstaining with DAPI in blue (4′,6-diamidino-2-phenylindole, SIGMA). The XY body is indicated (white arrow). After mounting the preparations with Vectashield (Vector Laboratories), the meiocytes were visualized using a Leica AFX6000LX multidimensional microscopy. The images were captured with LAS_AF software and analyzed and processed with public domain ImageJ and Adobe Photoshop CS3 softwares. All images are the result of superimposition of all the focal planes occupying the total volume of a pachytene mouse spermatocyte after squashing of seminiferous tubules.

We have not detected any ESCO2 labelling over the autosomas at any stage of meiosis. Since the X and Y chromosomes in all eutherian mammals preserve only a small region of homology, the genetic and morphological differentiation of sex chromosomes mark the XY behaviour during meiosis, which cannot be comparable with autosomes. Thus, structural modifications in the unsynapsed AEs of XY pair have been identified. Moreover, modifications in the distribution of different cohesin subunits have been also observed, as the preferential location of REC8 in the synapsed region of the XY body at pachytene (Page et al., 2006). However, our results show that the ESCO2 labelling embrace all the chromatin of sex chromosomes, similar signals have been detected after immunostaining of γ-H2AX and surprisingly the cohesin subunit RAD21L has been also observed in the XY chromatin at pachytene and diplotene (Herran, et al., 2011, Ishiguro, et al., 2011).

The presence of RAD21L over the XY chromatin has been explained as part of the sexual dimorphism observed regarding to the detection of RAD21L in mice and the authors pointed to a specific role of this meiotic cohesin subunit in the pairing and development of the sex body (Herran, et al., 2011), although deep studies should be performed to understand the role of a cohesin subunit in all sex chromatin. γ-H2AX is the histone variant derived from the phosphorylation of H2AX at serine 139 as consequence of DSBs (Mahadevaiah et al., 2001), however the presence of this chromatin modification in XY pair has been related to the transcriptional repression associated with the sex body (Fernandez-Capetillo et al., 2003).

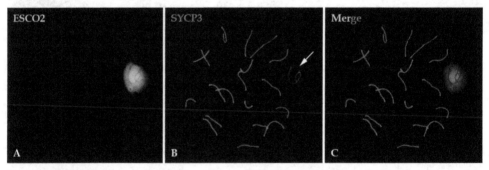

Fig. 3. Immunolabeling of ESCO2 (green) and SYCP3 (red) at pachytene after spreading of spermatocytes.
A. The immunolabeling of ESCO2 (rabbit anti-ESCO2 Bethyl Laboratories, A301-689A; 1:50 dilution) mark the XY body at pachytene. FITC-conjugated anti-rabbit IgG secondary antibody (Jackson Laboratories) was used at 1:150 dilution. **B.** The LEs of synpased autosomes and the AE/LEs of the sex chromosomes (white arrow) are detected with an anti-SCYP3 antibody (mouse monoclonal antibody anti-SYCP3. Santa Cruz Biotechnology, SCP-3

(D-1): sc-74569; 1:100 dilution). DyLight594-conjugated anti-mouse IgG secondary antibody (Jackson Laboratories) was used at 1:150 dilution. **C.** Merge image of ESCO2 and SYCP3 signals. After mounting the preparations with Vectashield (Vector Laboratories), the meiocytes were visualized using a Leica AFX6000LX multidimensional microscopy. The images were captured with LAS_AF software and analyzed and processed with public domain ImageJ and Adobe Photoshop CS3 softwares.

During male meiosis, X and Y chromosomes are silenced in a process called Meiotic Sex Chromosome Inactivation (MSCI), it has been described that this chromosome inactivation is essential for male meiosis progression, thus the disruption of MSCI arrest the male meiosis at pachytene, being essential for male fertility (Fernandez-Capetillo, et al., 2003, Royo et al., 2010). According to the formation of XY body and its transcriptionally silenced status during spermatogenesis different histone modifications have been described at pachytene such as H3 and H4 deacetylation or H3-K9 dymethylation (Khalil et al., 2004). The specific function of ESCO2 in the sex chromatin of male mice is not known. However it has been described that this acetyltransferase could play a role in regulating transcription, specifically it has been suggested a transcriptional repression activity through the interaction of ESCO2 with chromatin modifying enzymes and with the CoREST complex, achieving the repression by means of histone modification (Kim et al., 2008). Moreover, the authors talk about an ESCO2-containing complex, which has histone methyltransferase activity in culture cells. These evidences could link the presence of ESCO2 in sex chromatin and the MSCI process. Another cytological example of the specific presence of a cohesion cofactor in meiosis is related to WAPL localization. WAPL was found on axial and lateral elements of chromosomes (AE/LEs) in some prophase I stages in mouse spermatocytes (Kuroda et al., 2005) and oocytes (Zhang et al., 2008c) colocalizing with SYCP3, a constitutive protein of the SC; however, no more extensive study has been carried out on the role of WAPL in meiosis. It has been established that WAPL and PDS5 act as a protein complex during cohesion establishment and maintenance (see Skibbens, 2009 for further details); but there is no data about the possible presence of PDS5 in mammalian prophase I. However in budding yeast meiosis has been described that Pds5 is required for pairing of homologous chromosomes and this cofactor could have a role inhibiting the synapsis between sister chromatids. Depletion of Pds5 gives raise to fails in synapsis, hypercondense chromosomes and blocking meiocytes at pachytene-like stage (Jin et al., 2009). This result was preceded by the *Pds5* fission yeast mutant that showed absence of SC formation (Ding et al., 2006). The action of Pds5 in pairing and synapsis could be related with the interaction between both proteins and the Pds5 capacity of modulating Rec8 activity in both fission and budding yeast (Ding, et al., 2006, Jin, et al., 2009). Although these studies point out to the importance of the structural organization of meiotic chromosomes and the implication of cohesin complexes and cohesion cofactor in such important process as synapsis of homologous chromosomes, further cytological and biochemical studies are needed to characterize the role of PDS5 and WAPL during meiosis. Are those cohesion cofactors acting in unknown manner or their activity is related to the cohesion modulation? This is an unresolved question yet. Finally a recent molecular-genetic approach have shown that the cohesin-loading factor Scc2 and the subunits Scc3 and Smc1 of the cohesin complex are required for activating the production of the meiosis-specific subunit Rec8 in the *S. cerevisiae* (Lin et al., 2011). This result suggests that Scc2 could play a dual role in gene regulation and sister-chromatid cohesion during meiotic differentiation. Although the mechanism is poorly understood yet, the role of cohesins and cohesion cofactors in gene regulation has been largely studied in the somatic line (Chien et al., 2011, Dorsett, 2011,

Newman & Young, 2010, Ong & Corces, 2011). However, meiosis-specific gene regulation is a future field of study where probably different groups are work in.

5. To form bivalents: Cohesion, synapsis and recombination

Two of the most important phenomena that characterize the meiosis are the SC formation and the reciprocal homologous recombination. Both are meiosis-specific and are implicated in the physical connexion of homologous chromosomes to form and maintain the bivalents. In this context, the cohesion is the third pivotal factor. Since the cohesion stabilizes the joint between sister chromatids from S phase until their segregation, it is easy to think that the cohesin complexes are needed to ensure that both sister chromatids of each homologous chromosome act in a single manner during prophase I, in a cellular context where the AEs formation, the synapsis and the meiotic recombination between homologous chromosomes is taking place. This would be the role of cohesin complexes in sister chromatid cohesion, its canonical function. Thus, during first meiotic division these three events: cohesion, synapsis and recombination, must be perfectly coordinated. However, the existence of meiotic-specific cohesins and its specific localization during first meiotic division become evident that the meiotic cohesin complexes could be implicated in SC formation and recombination directly further than the canonical function of chromatid cohesion.

5.1 The importance of individual cohesin subunits to synapsis

The SC is a meiosis-specific proteinaceous structure formed by two lateral elements (LEs), derivate from AEs of chromosomes, and a central element (CE) connected by transverse filaments (TFs). It is known that this meiotic structure play important roles in the condensation and pairing of chromosomes and also has a close relationship to the programmed DSBs formation and resolution (Page & Hawley, 2004). The formation of SC between homologous chromosomes is highly conserved in evolution, it derivates in the synapsis of homologous chromosome during the meiotic prophase I. One of the first references of the close relationship between the SC and cohesion was observed in budding yeast (Klein et al., 1999). In these organisms the cohesin subunit Smc3p presents a continuous localization along chromosome cores similar to that described for SC proteins in rat (Schalk et al., 1998). Klein et al., included a functional analysis, they studied the *Smc3* and *Rec8* mutants and observed that both were defective in cohesion, formation of AEs and SC assembly, demonstrating in yeast that the cohesins were essential not only for the meiotic sister chromatid cohesion but also for the synapsis. Moreover they postulated that the meiotic cohesin complexes were in fact part of AEs of chromosomes. However, two years later Pelttari et al., went farther and proposed that the cohesin axis was preformed along each chromosome and might act as the organizing framework for AE/LEs formation. This proposal was based on the evidences that *Sycp3-/-* mice were able to form a cohesin axis even in the absence of AEs (Pelttari et al., 2001). Contrary, the *Rec8-/-* and *Smc1β-/-* mice showed AEs partially assembled (Revenkova, et al., 2004, Xu et al., 2005). In rat spermatocytes the interaction between SMC cohesin subunits and SYCP2 and SYCP3, the main components of AEs and LEs, was also studied, concluding that the cohesin axes were essential in the SC formation and synapsis progression during early stages of prophase I (Eijpe et al., 2000, Eijpe, et al., 2003). These first observations pointed to the contribution of cohesins not only to form the AEs but also as part of AEs/LEs and then to its function in synapsis. Since then we have known different examples which closely relate the SC proteins and the synapsis progression to cohesin localization and function in meiosis. In the table 1 we summarize some of examples of the contribution of cohesin to SC formation and synapsis that we detail forward.

Organism	Cohesin	Role in AE/LEs formation or synapsis	Reference
C. elegans	REC-8	Chromosome pairing AE component SC assembly	Chan 2003 Colaiacovo 2006 Pasierbek 2001
S. cerevisiae	SMC3p REC 8	AE/LEs localization SC assembly	Klein 1999
Rat	SMC1α/β SMC3	AE/LEs formation Interaction with SYCP2 and SYCP3	Eijpe 2000 Revenkova 2001
	REC8	AEs assembly	Eijpe 2003
	SMC1α SMC3	Part of AEs	
Tomato	SMC3 SMC1 SCC3 REC8	AE/LEs localization Sequential assembly	Qiao 2011 Stack 2009
Grasshoppers	SMC3 SA1 RAD21	AE/LEs localization Present in synapsed regions exclusively	Valdeolmillos 2007
Mice	STAG3	Association to SC	Prieto 2001
	REC8	Homologous SC assembly	Xu 2005
	RAD21L	AEs formation and synapsis	Herrán 2011
	SMC1α	Synapsis	James 2002
	SMC1β	AEs formation and SC assembly	Revenkova 2004

Table 1. Role of different cohesin subunits in AE/LEs formation and synapsis

The evidences of the implications of cohesin complexes in SC assembly, pairing and/or AEs formation came from many different organisms and scientific approaches. In plants, a specific and intermittently localization of SMC3 in the AE/LEs has been observed in tomato by means immunogold labeling and electron microscopy in zygotene microsporocytes (Qiao et al., 2011, Stack & Anderson, 2009), similar to that observed by light microscopy after the immunolabeling of SMC1, SMC3, SCC3 and REC8, although no all subunits presented the same pattern of accumulation and appearance during prophase I (Qiao, et al., 2011). In this vegetal species REC8, SMC1 and SMC3 are localized as foci from preleptotene which are arranged to AEs at leptotene but do not colocalize. However they exhibit colocalization in synapsed regions at zygotene and pachytene. SCC3 appears later than the other three cohesin subunits, observing a substantial amount of SCC3 foci along AE/LEs from zygotene through pachytene, according with the SC assembly. There is no a functional analysis about the implication of this vegetal cohesin in SC formation and synapsis, but this sequential loading shows a possible change of cohesin complexes composition or organization at a meiotic stage which is key in the correct progression of synapsis . This non-homogenous timing and spatial localization of different cohesin proteins has been reported in other organisms such as *C. elegans* (Chan et al., 2003) or male grasshoppers (Valdeolmillos et al., 2007). Males of grasshoppers *Locusta migratoria* and *Eyprepocnemis plorans* species represent another cytological example of the contribution of cohesins to synapsis. In grasshoppers the SMC3 subunit appeared as a dotted signal in preleptotene to form continuous lines in leptotene, meanwhile the non-SMC subunits RAD21 and SA1 were not localized until zygotene and only on synapsed regions. This means that in grasshoppers the incorporation of non-SMC cohesins would be spatially and temporally concomitant with the progression of synapsis. In zygotene and pachytene SMC3 and RAD21 present a continuous and linear pattern, similar to what would be expected of AE/LEs proteins and both cohesin subunits present a complete colocalization until diplotene, coinciding with the stages of disassemble of SC. It is relevant that the single X chromosome of grasshopper males, which remains unsynapsed, presents neither RAD21 nor SA1 during prophase I. Further than the possible implication of RAD21 and SA1 in the complete assembly of SC or *vice-versa*, these observations open new questions: Would this synapsis-dependent sequential loading of cohesin during prophase I have an effect on the distribution of cohesins in metaphase I? Could we talk about a presence of cohesin depending on the synaptic history of chromosomes? At any case, these kind of studies pointed not only that cohesin subunits can be loaded onto meiotic chromosomes at different time or as part of different complexes, but also relate the loading of specific cohesin subunits with the timing of AEs formation and SC assembly. The idea of specific synapsis cohesin complexes is also supported by the evidence that SMC1α is mainly located along synapsed regions in mice spermatocytes, whereas SMC3 subunit localizes along the synapsed and unsynased AEs (James et al., 2002), in front of SMC1β which is mainly required for sister chromatid cohesion (Revenkova, et al., 2004). In addition, *Sycp3-/-* and *Smc1β-/-* single and double oocytes mice mutants show that both proteins contribute to meiotic chromosome axis organization (Novak et al., 2008). The authors propose a several layers axis association, which contributed to the formation of meiotic chromosome axes in an independent manner. The basic layer would be formed by the same proteins that form the mitotic chromosome axis, the scaffold proteins (Maeshima & Laemmli, 2003). SMC1β would be part of the second layer, it contributes to axes length but is not essential (Revenkova, et al., 2004). Finally SYCP3 would represent the third layer as basic component of AE/LEs. Thus a cohesin axis would be defined as a base axis before the AE/LEs in mammals, a similar organization has been proposed in grasshoppers (Valdeolmillos, et al., 2007). *Smc1β -/-* mice spermatocytes do not progress further than mid-pachytene and the cells present abnormalities in chromosome structure (Revenkova, et al., 2004). The analysis of SC proteins showed that the AEs were shortened, but the synapsis occurred between the 19 couples of homologous chromosomes. Thus, despite of the close relationship of the cohesins and SC proteins and the evident interdependence between both groups of proteins, there are examples in mice where the absence of a cohesin subunit does not inhibit the complete formation of SC, albeit its structure is compromised (Revenkova, et al., 2004, Xu, et al., 2005). This could be explaining due to the presence of different cohesin complexes in mammal meiotic cells (Revenkova & Jessberger, 2005, Revenkova & Jessberger, 2006, Suja & Barbero, 2009) which could act in a redundancy manner. All these evidences suggest that the cohesins and specifically the SMC cohesin provide a basis for AE assembly. Based on the description of the new cohesin subunits and regarding to its role in SC formation and synapsis, diverse authors have develop models of incorporation of new cohesin complexes or replacement of specific cohesin subunits during the early prophase I that we summarize in figure 4.

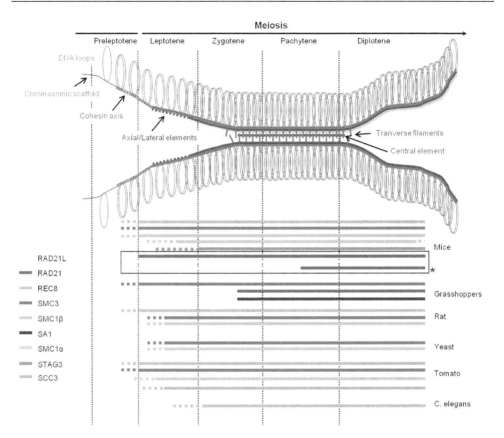

Fig. 4. Temporal meiotic appearance of cohesin subunits from preleptotene and its relationship with SC formation.

At the top of the figure a temporal line of early stages of meiosis progression is shown. The diagram positioned under this line is represented the cohesin axis formation and the SC assembly taking into account the data observed in mice. The blue lines represent the chromosomic scaffold and the blue circles are the chromatin loops. The pink color represents the cohesin organization, which is usually detected as a dotted pattern at preleptotene and from leptotene form axis that would the base for the AEs proteins. The AE/LEs are represented in green, the FTs in red and the CE of SC in purple. At the bottom of the figure, under the diagram of SC assembly, the temporal localization of different cohesin subunits in different species is shown. The colour code for each cohesin subunit is situated at the left.

* This box shows the disagreement regarding to the RAD21 and RAD21L dynamic. See text for further detail.

In the nematode *C. elegans*, REC-8 and HIM-3, a component of meiotic chromosomes axes required for synapsis (Zetka et al., 1999), overlap during leptotene, zygotene and pachytene, suggesting that REC-8 is a component of AE/LEs (Pasierbek et al., 2001). SMC-1 and SMC-3 loaded onto chromatin independently of REC-8 and SCC-3. On the other hand the depletion of *Rec-8* by RNAi did not cause change in SMC-1 or SMC-3 localization but a mislocalization of SCC-3 was detected (Chan, et al., 2003). The importance of REC-8 as an associated component of AEs in *C. elegans* derives from the evidence that this protein is presented along the axes of unsynapsed chromosomes and this meiosis-specific cohesin subunit seems to be necessary for the assembly of SC components such as HIM-3, SYP-1 and SYP-2 (Colaiacovo, 2006, MacQueen et al., 2002, Nabeshima et al., 2005, Pasierbek, et al., 2001). Moreover, the analysis of *Rec-8* (RNAi) mutants showed that this protein is implicated in the chromosome pairing and SC assembly during leptotene/zygotene stage and the meiotic chromosomes are reorganized before diplotene into pachytene-like arrangement but they fail to achieve a normal lengthwise alignment (Chan, et al., 2003). Probably REC8 is the subunit of cohesin more widely studied in meiosis, it is a meiotic-specific cohesin variant of the subunit SCC1/RAD21. REC8 has orthologs in different organisms, it is present in mammals (Xu, et al., 2005), yeast (Klein, et al., 1999, Molnar, et al., 1995), plants (Qiao, et al., 2011, Zhang et al., 2006) and *C. elegans* (Pasierbek, et al., 2001). Although it has not been defined an obvious REC8 homologue in *Drosophila melanogaster*, C(2)M is a protein distantly related to REC8 which has not play a cohesion role but localized to LEs and interacts with SMC3 (Anderson et al., 2005). In rat spermatocytes the meiosis-specific REC8 protein was observed as axial structures during premeiotic S phase, earlier than SMC1β and SMC3 which appeared during leptotene along with SYCP2 and SYCP3, suggesting an special role of REC8 in AEs formation (Eijpe, et al., 2003, Revenkova, et al., 2001). The meiocytes of males and females *Rec8* null mice do not complete the prophase I and presents SC assembly between sister chromatids (Xu, et al., 2005). Thus, in mammals REC8 seems to be necessary to the formation of SC between homologous chromosomes instead of the properly assembly of SC, albeit its structure is altered. Recently a new meiosis-specific SCC1 cohesin subunit has been identified in vertebrates, it is closely related to RAD21 and it has been named RAD21L (RAD21-like) (Gutierrez-Caballero, et al., 2011, Ishiguro, et al., 2011, Lee & Hirano, 2011). RAD21L interacts with other cohesin subunits such as SMC1α/β, SMC3 or STAG3 but does not interact with REC8, thus both kleisins subunits form distinct meiosis-specific cohesin complexes (Gutierrez-Caballero, et al., 2011, Ishiguro, et al., 2011). There is no agreement regarding to the localization of RAD21L from mid-pachytene/diplotene, thus meanwhile some authors ensure that RAD21L disappears from chromosomes axes at mid-pachytene when the SC is still formed (Lee & Hirano, 2011), others have found that RAD21L is dissociated from AEs on autosomes in a progressive manner, during SC disassembly at diplotene (Herran, et al., 2011, Ishiguro, et al., 2011) and it is accumulated at centromeres until anaphase II (Herran, et al., 2011). Similarly, there are different results in relation with the coexistence of RAD21 and RAD21L complexes during early prophase I stages. Lee and Hirano (2011) ensure that RAD21L seems to replace RAD21 from leptotene, finding that RAD21L localizes along AE/LEs until pachytene, accompanying the SC formation and homologous synapsis. However, RAD21 has been previously detected by immunolocalization during leptotene, zygotene and pachytene in mouse spermatocytes by other authors (Parra, et al., 2004, Xu et al., 2004) and at least in the first case, the antibody used in Parra's et al., study does not cross-react with RAD21L as it is shown in Herran's et al., paper., suggesting the presence of both RAD21 and RAD21L at the same time. At any case REC8, RAD21 and RAD21L are detected in a mutually exclusive manner, forming part of different cohesin complexes. The definitive evidence to ensure that RAD21L is directly implicated in AEs formation and synapsis is the *Rad21L-/-* male mouse which exhibits discontinuous AEs stretches and fails in homologous synapsis, moreover the mutants present defects in DSBs processing (Herran, et al., 2011). The RAD21L knockout males are sterile but females are fertile although develop age-dependent sterility is observed. On the other hand, there are evidences of the interaction of RAD21L and SYCP1, main component of TFs of SC (Meuwissen et al., 1992), and the centromere protein CENP-C (Ishiguro, et al., 2011, Lee & Hirano, 2011). In a general manner, there is a rising common idea about the specific implication of RAD21L in pairing and the initiation of synapsis between homologous chromosomes (Herran, et al., 2011, Ishiguro, et al., 2011, Lee & Hirano, 2011, Polakova et al., 2011). However, many questions are still open and new interesting doubts have been joined the fray regarding to the regulation of RAD21L cohesin complexes (Uhlmann, 2011).

In mouse testis the meiotic protein STAG3 was first identified as a stromalin member associated to SC (Pezzi, et al., 2000) and after that was described its participation in the cohesin complex and its specificity to cohesion of sister chromatid arms during first meiotic division (Prieto, et al., 2001), however the authors pointed that STAG3 could be a component of SC, being detectable on chromatin paralleling to AEs appearance. This is an interesting example because STAG3 is the fourth component of the cohesin complex, it is not a kleisin and does not interact with SMC proteins which might be the base of the cohesin axes formed during the early stage of prophase I. Thus in a general model it has been observed a correlation between the progression of AE/LEs formation and synapsis and the localization of several cohesin subunits. According to the studied organisms the localization of cohesin subunits during prophase I has been observed continuous or discontinuous; however this could be related with the preparation method or the meiotic stage. In any case, all cohesin subunit are associated with AE/LEs during prophase I. What we consider more relevant is the sequential loading of cohesins observed during the first stages of prophase I and how the appearance of some subunits seems to be closely related with the progression of AE/LEs and SC formation and synapsis progression. In this sense, it is important to consider if the association of the cohesin subunits with the AE/LEs could be in an independent manner one to another. These cases would be possible only if we consider that in these species different cohesin complexes exist, as different author postulated before (Chan, et al., 2003, Eijpe, et al., 2003, Qiao, et al., 2011, Suja & Barbero, 2009, Valdeolmillos, et al., 2007), being some of them more intimately related with the progression of synapsis or the SC formation. But then, we should think about if the cohesins loaded during early prophase I are really exerted a cohesin function or are related with AE/LEs formation and/or synapsis progression. Different analysis in mitotic cells showed that only the cohesins loaded during S phase are effective in the cohesion of sister chromatids (Austin et al., 2009, Guacci, 2007), despite of evidences about new cohesion at DBSs point (Strom & Sjogren, 2005). Generally, the SMC cohesins seems to be more closely related with the AE/LEs formation, derivates from the evidences that this cohesin subunits are the base of the cohesin complexes then they would be the base of a cohesin axis and appear during the stages where SYCP2 and SYCP3 are incorporated or a bit early (Eijpe, et al., 2000, Klein, et al., 1999, Novak, et al., 2008, Qiao, et al., 2011, Revenkova, et al., 2001, Valdeolmillos, et al., 2007). However, the non-SMC proteins seem to have more specific roles before AEs formation (Eijpe, et al., 2003, Gutierrez-Caballero, et al., 2011, Ishiguro, et al., 2011, Molnar, et al., 1995) or during SC assembly (Valdeolmillos, et al., 2007). Especially we are thinking in the SCC1 subunits and all its variants, and then a cohesin complex would acquire a specific role depending on the SCC1 which is formed with. It is difficult to split out the study of cohesin distribution during first meiotic division and the function of SC as the physical connector of homologous chromosomes at prophase I. More and more the scientists notice that the cohesins present function or contribute to function that historically had been attributed to the SC or to specific SC proteins and different cohesin complexes acquire specific functions in meiosis.

5.2 Cohesins and meiotic recombination

The first evidence that sister chromatid cohesion is required for DNA damage repair was the observation that the expression of mutated cohesins or the deletion of some subunits cause hypersensitivity and fails in DNA repair after induced or spontaneous DSBs in somatic line (Birkenbihl and Subramani, 1992; Sonoda et al., 2001). Cohesin complexes are preferred loaded at intergenic sites (Lengronne et al., 2004) onto the replicating chromosomes during S phase. However, now it is clear that in somatic cells the DNA damage induces the recruitment of cohesins after replication and lead to the formation of new cohesion (Strom and Sjogren, 2005), which is essential for DNA repair (Strom et al., 2004). Since the ability of SMC proteins to promote reannealing of complementary DNA strand was described, a role of cohesins in resolving DSBs was suggested (Jessberger et al., 1996). Different authors have studied and reviewed the role of cohesins in DSBs repair in S and G2 periods in mitotic cells (Cortes-Ledesma & Aguilera, 2006; Nagao et al., 2004; Strom & Sjogren, 2007; Watrin & Peters, 2006), however, during meiosis the DSBs are programmed, in front of the spontaneous breaks that can alter the genome during the rest of the cell cycle. Basically mitotic/somatic and meiotic recombination have different purposes, thus whereas the meiotic recombination is initiated to generate a crossover (CO) on each bivalent at least, the mitotic recombination is used to repair DNA damage to ensure the fidelity of DNA sequence and cell surveillance. This important difference affects the repair pathway (Andersen & Sekelsky, 2010). We do not pretend to review all the literature regarding to the function of cohesin in all kinds of DNA damage, we will focus our attention in those cohesin subunits that have a specific role in meiotic DSBs and recombination.

Most of current models of meiotic recombination are based in the model that Szostak and colleagues proposed in yeast in 1983. This model is based on the generation of DSBs at early prophase I, the strand invasion at zygotene, the formation of double Holliday junction (dHJ) intermediate and its resolution at pachytene/diplotene. Obviously the repair mechanisms of DNA molecules act in an organized chromosome, where protein axial structures such as AEs and cohesin axis exist. The DSBs may be resolved toward CO between homologous chromatids or NCO (non-crossover), this decision can be made at different steps or repair pathway (see Heyer et al., 2010 for further details). Meiotic COs are essential for ensuring chromosome segregation. Chiasmata, points of genetic interchange between non-sister chromatids, keep the homologous chromosomes as a single structure after the disassembly of SC. Chiasmata and sister chromatid cohesion ensure the correct orientation of bivalents at metaphase I. Thus, in a similar way that fails in cohesion lead premature separation of chromatids, fails in meiotic recombination promote the generation of univalents at metaphase I. In both cases a high probability of aneuploidy exists. Meiotic recombination is initiated after a replication period, thus the sister chromatids are linked by cohesins when the recombination process begins. It has been established that cohesins are needed for the formation and stabilization of DNA interconnections (Buonomo et al., 2000; Klein et al., 1999).

Meiotic recombination begins via DSBs formation by the Topoisomerase II-like enzyme SPO11 (Grelon et al., 2001; Keeney et al., 1997). As consequence of SPO11 activity a great amount of DSBs are generated during early meiotic prophase I and most of them are repaired as NCO. Thus the cohesins take part in bivalent stabilization and recombination to ensure that the excess of programmed DSBs do not alter the chromosome organization. On the other hand specific roles of different cohesin subunits in DSBs formation and regulation have been also described. As we stated above the basic difference between meiotic and mitotic recombination lies in the generation of DSBs, thus SPO11 has a pivotal role in the activation of meiotic recombination machinery and the regulation of this enzyme is one of the most interesting topics of meiotic recombination analysis. SPO11 has the key to begin all the process of genetic exchange between non-homologous chromosomes. Different studies have shown a non-random distribution of meiotic DSBs (Baudat & Nicolas, 1997; de Massy, 2003; de Massy et al., 1995; Gerton et al., 2000), giving raise to hot and cold-spots recombination sites. Different factors, as some histone variants (Buard et al., 2009) or the DNA sequence (Grey et al., 2009), determine and regulate the action of SPO11. In this context the cohesins have also been implicated. In budding yeast the initial centromeric entry of SPO11 depends on REC8 and this subunit of cohesin complex choreographs the distribution of SPO11 to DSBs sites (Kugou et al., 2009). On the other hand, in *S. pombe*, the *Rec8Δ* mutant shows a marked reduction of meiosis-specific DNA breakage by REC12 (the *S. pombe* SPO11 homologue) (Ellermeier & Smith, 2005) and in *C. elegans* has been proposed that REC-8 could be implicated in limiting the activity of SPO-11 (Hayashi et al., 2007). However, whereas in *S. cerevisiae Rec8* mutants the chromosome segregation is at random (Klein et al., 1999), similar mutants in *S. pombe* present a equational segregation of sister chromatids at anaphase I (Watanabe & Nurse, 1999). These discordant results suggest that the regulation of meiotic cohesin complexes containing REC8 and its role in recombination depends on the species. After the formation of DSBs by SPO11, modifications on chromatin as the phosphorylation of histone variants H2AX, H2Av and H2B occur (Fernández-Capetillo et al., 2004; Madigan et al., 2002; Redon et al., 2002) and a cascade of repair proteins are loaded onto damage sites in order to restore the DNA properly. The first recombinases that bind to chromatin after DSBs are RAD51 and DMC1. The last one is a meiosis-specific protein. Usually the recruitment of recombinases has been associated to the AEs (Plug et al., 1998; Tarsounas et al., 1999), but the cohesin axis is also functional in recruiting DNA recombination proteins. This issue was observed in *Sycp3-/-* mice spermatocytes, where the recombinase RAD51 is recruited in the absence of AEs (Yuan et al., 2000) suggesting a higher level of organization which is probably related to the cohesins. The resection of single-strand DNA allows the single-end invasion promoted by both recombinases RAD51 and DMC1 (Neale and Keeney, 2006; Shinohara and Shinohara, 2004). The extended invading strands are then processed to form CO or NCO. At this point the most of DBSs are resolved by Synthesis-Dependent Strand Annealing (SDSA) through NCO (Allers & Lichten, 2001). Some points of strand invasion are stabilized to form a D loop and double Holliday junctions (dHJ) intermediate, which can be resolved through CO or NCO depending on the cleavage of dHJ. The DSBs formation and the CO/NCO decision are two relevant items in meiotic recombination studies. In mice, before the AEs formation a cohesin core is observed as we reviewed above, but the action of SPO11 is not required for the formation of cohesin cores as it was observed in *Spo11-/-* spermatocytes, being SMC1α mostly associated to homologous or non-homologous synapsed regions meanwhile SMC3 localized both to AEs of unsynapsed and synapsed cores at arrested zygotene cells (James et al., 2002). In mammals, the absence of REC8 does not affect the timing of DSBs formation and the recruitment of the recombinases RAD51/DMC1, suggesting than DSBs and the early recombination events occur properly. But, the persistence of RAD51/DMC1 foci in *Rec8-/-* mice spermatocytes and the absence of MLH1 foci might reflect the inability to repair the DSBs (Xu et al., 2005). MLH1 and MLH3 protein are implicated in resolution of recombination to chiasma (Hunter & Borts, 1997; Wang et al., 1999). It is not clear if REC8 has a specific role in resolving the meiotic DNA damage or the persistence of unresolved DSBs is a consequence of that the *Rec8-/-* cells do not reach the meiotic stage where the recombination resolution takes place. In this rodent REC8 seems to have a pivotal role in sister chromatid cohesion and synapsis but its relationship with recombination is not clear, despite of a small proportion of REC8 coprecipitate with RAD51/DMC1. Moreover a small proportion of RAD50 coimmunoprecipitate with REC8 from spermatocyte lysates but REC8 does not coprecipitate with RAD50 (Eijpe et al., 2003). RAD50, MRE11 and NBS1 proteins form the MRN complex which is implicated in the early response to DNA damage. The action of MRN complex is related to nuclease activity at the 5' ends, releasing SPO11 from DNA (Lichten, 2001). After the action of MRN, RAD51 and DMC1 can bind to single-strand DNA. These results which relate REC8 with recombination repair proteins would indicate that protein complexes that contain both REC8 and RAD51/DMC1 and some RAD50 exist in spermatocytes (Eijpe et al., 2003). The authors proposed that after S phase, cohesins would attract protein complexes that are involved in the early steps of homologous recombination. Recent results in yeast proposed that Rec8 could have a role in maintaining the homolog bias of recombination (Kim et al., 2010). In *Rad21L-/-* mice spermatocytes an accumulation of unrepaired DSBs has been described, but the cells are arrested at zygotene-like stage (Herran et al., 2011). Although, in sight of the cytological results of appearance and disappearance of RAD21L in wild type mice, a role in meiotic recombination has been suggested for this cohesin subunit (Herran et al., 2011; Ishiguro et al., 2011), deeply studies are needed to corroborate the implication of RAD21L in the processing of recombination intermediates. However, in this context we might probably talk about a role of SMC1β in recombination. Initially this protein was found at bridges between homologous AEs, and it was supposed that these represented the chiasmata sites (Revenkova et al., 2001). In *Smc1β-/-* spermatocytes early markers of recombination as RAD51 and γ-H2AX localize properly. The cells progress through mid-pachytene and MLH1 and MLH3 foci were not detected in males, additionally the *Smc1β-/-* oocytes develop further and also fail to form the normal number of MLH1 foci (Revenkova et al., 2004). Probably the effects observing in cohesin mutants regarding to recombination proteins could be explain due to improper sister chromatid cohesion, an incorrect formation of AEs or synaptic fails. Perhaps we have to analyze the relationship between cohesin and meiotic recombination in a cohesin-dependent chromatin organization. An example is found in *S. pombe*, in this organism Rec8 and Rec11, meiosis-specific cohesin subunits, are essential for DSBs only in some regions of the genome (Ellermeier & Smith, 2005). On the other hand the loading of Rec25 and Rec27, two recombination proteins in yeast, depends on the previous incorporation of Rec8 (Davis et al., 2008). This issue could be explained because of Rec8 is required for the normal chromosome compaction during meiotic prophase I (Ding et al., 2006), thus it has been proposed that the regional effect on DSBs production might be related to the meiotic chromosomal organization formed by the cohesins (Davis et al., 2008). Thus the role of cohesins in meiotic recombination should be at different levels, chromosomal organization, providing the proximity between sister chromatid to ensure a correct repair with an undamaged template and as individual proteins that form complexes with recombination proteins.

Finally we would like to comment two interesting cytological examples of the relationship between cohesin axis development, synapsis and recombination. The males of *Stethophyma grossum* exhibit a delay in cohesin axis formation in the eight longest bivalents of its complement and in the single X chromosome (Calvente et al., 2005). These regions remain unsynapsed, presenting three different synaptic situations: an unsynapsed sex chromosome, three bivalents with complete synapsis and eight bivalents with incomplete synapsis (Jones, 1973; Wallace & Jones, 1978). In this species the DSBs and the recruitment of first recombinases occur before the beginning of synapsis, detecting RAD51 foci over the cohesin axis at leptotene. But the most interesting event is that γ-H2AX and RAD51 foci are polarized to the nuclear region where the cohesin axis are formed in advanced and these regions correspond with the ulterior synapsed chromosomal regions. Thus, those chromosomal regions that at leptotene do not exhibit a formed cohesin axes do not present DSBs and then do not recruit RAD51 (Calvente et al., 2005). A similar case has been observed in *Paratettix meriodinalis*, another grasshopper species with an incomplete synapsis pattern. The spermatocytes of this species show a polarization of cohesin axes maturation and the initiation of meiotic recombination, suggesting that the early development of cohesin axis would drive the loading of recombination machinery (Viera et al., 2009b). Obviously the study of AEs proteins would clarify the real relationship among SC formation, cohesin axis development and recombination in this special organism but these examples show the relevance of a cohesin axis formed by at least SMC proteins not only in the beginning of meiotic recombination but also in synapsis progression.

6. Cohesins and cohesin-regulators in human cancer and cohesinophaties

As we commented before, dysfunction of chromosome cohesion during meiosis frequently gives fertility problems. However, now we know that cohesins participate in other important cell processes, such as mitotic DNA damage repair and control of gene expression. One of the most surprising findings was the discovery of a link between cohesin mutations and human diseases. Two groups, (Krantz et al., 2004, Tonkin et al., 2004), found that the *Nipbl* gene, the homologue to yeast *Scc2* adherin gene, is mutated in the human CdLS (OMIM: 122470, 300590, 610759). This pathology is a multiple neuro-developmental disorder characterized by facial dysmorphisms, mental retardation, growth delay, and upper limb abnormalities. Subsequent studies of several cases of this syndrome showed that mutations in SMC1α and SMC3 (Musio et al., 2006, and Deardorff et al., 2007) cause a mild variant of CdLS. Roberts syndrome/SC phocomelia (RBS; OMIM: 268300) is an autosomal recessive disorder phenotypically related to CdLS; RBS patients present craniofacial abnormalities, growth retardation, and limb reduction. Cells from RBS patients show a lack of cohesion at the heterochromatic regions around the centromeres and at the Y chromosome long arm (Van Den Verg and Francke, 1993). It has been described that RBS is caused by mutations in *ESCO2* gene (Vega et al., 2005) and analyses of different mutations in RBS patients pointed to the loss of ESCO2 acetyltransferase activity as the molecular mechanism involved in RBS pathology (Gordillo et al., 2008). On the other hand, in the last years have been published different studies and experimental data that links cancer development with disorders in the core cohesin subunit and cohesin- interacting proteins genes. Although, we do not mind to review this field here, the following paragraphs reveal the importance of cohesins, including meiotic-specific cohesin genes in tumorogenesis. The study of 11 somatic mutations in 132 human colorectal cancers identified 6 of them mapping to 3 cohesin, *SMC1a*, *SMC3* and *STAG3*, genes and 4 to a cohesin-regulator *SCC2* gene (Barber et al., 2008). Colorectal cancer cells are characterized by chromosomal instability, resulting in chromosome gain or loss. It is possible to argue that abnormal cohesin pathway activity leads to chromosome missegregation and chromosome instability. This hypothesis is supported by the observation that colorectal cancer cells exhibit up to 100-fold higher rates of missegregation than normal cells (Lengauer et al., 1997). In adittion, using a microcell mediated chromosome transfer and expression microarray analysis the cohesin *STAG3* gene was identified as one of the nine genes associated with functional suppression of tumorogenicity in ovarian cancer cell lines (*AIFM2, AKTIP, AXIN2, CASP5, FILIP1L, RBBP8, RGC32, RUVBL1* and *STAG3*) and as a candidate gene associated with risk and development of epithelial ovarian cancer (Notaridou et al., 2011). The future research on the molecular mechanisms surrounding the cell regulation of cohesin is crucial to understand the relationship between the cohesin-interacting proteins and cohesin post-translational modification with developmental alterations and cancer. These studies probably will help to understand the relationships between this interesting ring protein complex and the formation/development of tumors in humans.

7. Conclusions

In the late 90's, cohesins were discovered and characterized are the main actors in controlling the sister chromatid joint until the chromosomic segregation. This important issue ensures a correct distribution of chromosomes to daughter cells and then the viability and continuity of all organisms. Undoubtedly the cohesion regulation leads the chromosomic segregation in mitosis and meiosis. However, whereas a correct regulation of cohesion in mitosis is crucial for development, in meiosis this process is vital for the fertility of the individuals. The complexity of meiosis arises from the necessity of generate haploids cells and the magic of this type of cell division is that all the daughters cells are different among them and to the mother cell. In order to carry out this great objective, the homologous chromosomes form bivalents, recombine, organize a complex protein structure between them, the SC, and go through two consecutive segregations. In this context the cohesion and the cohesin complex in meiosis present different specializations such as meiosis-specific cohesin subunits or a strict regulation of the release of arm and centromeric cohesion during both chromosomic segregations. As an example of this complexity, nowadays four meiosis-specific cohesin subunits have been identified in mice: SMC1β, STAG3, REC8 and RAD21L. These subunits do not exclude the presence of other mitotic proteins, forming then a high number of different cohesin complexes. However, further than the role of cohesin complex in cohesion, all the specializations observed in meiosis seem to have an effect on functions of the cohesins. In this scenario we are now discovering the action of this fascinating group of proteins in the formation of AE/LEs, the progression of synapsis, the formation of programmed DSBs and the regulation of gene expression in meiosis. In a same way, we begin to understand the role of cohesins in developmental diseases and cancer. It have been identifying a growing number of proteins that interact with the cohesin complex and modulate the dynamics of cohesins and its binding to chromatin. Since the cohesins do not act in a single manner but they need different proteins to regulate their action, the cohesins-regulators have reached the attention of the scientific community. In addition, today we know that mutations in genes encoding cohesin regulators could give pathological conditions in humans. Update we do not know meiosis-specific regulators, but the action of the cohesin-interacting proteins must be regulated in a meiosis-specific manner and probably there are meiotic factors, which are waiting to be discovered. Many doubts are still unresolved and many doubts will appear later, but at any case we cannot study the cohesion and the cohesins as a single concept but a complex process and a versatile group of proteins which are implicated in the correct progress of the cell cycle in a global manner.

8. Acknowledgments

We thank Dr. Mónica Pradillo for her sincere and critical comments. We apologize to all colleagues whose important contributions have not been referenced due to space restrictions. This work was supported by the Spanish Ministerio de Ciencia e Innovación (grant BFU2009-08975/BMC) and CSIC (grant PIE-201120E020).

9. References

Allers, T. & Lichten, M. (2001). Differential timing and control of noncrossover and crossover recombination during meiosis. *Cell*. Vol.106, No.1, pp.47-57, ISSN:0092-8674

Andersen, S.L. & Sekelsky, J. (2010). Meiotic versus mitotic recombination: two different routes for double-strand break repair: the different functions of meiotic versus mitotic DSB repair are reflected in different pathway usage and different outcomes. *Bioessays*. Vol.32, No.12, pp.1058-1066, ISSN:1521-1878

Anderson, L.K., Royer, S.M., Page, S.L., McKim, K.S., Lai, A., Lilly, M.A. & Hawley, R.S. (2005). Juxtaposition of C(2)M and the transverse filament protein C(3)G within the central region of Drosophila synaptonemal complex. *Proc Natl Acad Sci U S A*. Vol.102, No.12, pp.4482-4487, ISSN:0027-8424

Austin, C., Novikova, N., Guacci, V. & Bellini, M. (2009). Lampbrush chromosomes enable study of cohesin dynamics. *Chromosome Res*. Vol.17, No.2, pp.165-184, ISSN:1573-6849

Barber, T.D., McManus, K., Yuen, K.W., Reis, M., Parmigiani, G., Shen, D., Barrett, I., Nouhi, Y., Spencer, F., Markowitz, S., Velculescu, V.E., Kinzler, K.W., Vogelstein, B., Lengauer, C. & Hieter, P. (2008). Chromatid cohesion defects may underlie

chromosome instability in human colorectal cancers. *Proc Natl Acad Sci U S A.*
Vol.105, No.9, pp.3443-3448, ISSN:1091-6490

Baudat, F. & Nicolas, A. (1997). Clustering of meiotic double-strand breaks on yeast
chromosome III. *Proc Natl Acad Sci U S A.* Vol.94, No.10, pp.5213-5218, ISSN:0027-
8424

Beckouet, F., Hu, B., Roig, M.B., Sutani, T., Komata, M., Uluocak, P., Katis, V.L., Shirahige,
K. & Nasmyth, K. (2010). An Smc3 acetylation cycle is essential for establishment of
sister chromatid cohesion. *Mol Cell.* Vol.39, No.5, pp.689-699, ISSN:1097-4164

Birkenbihl, R.P. & Subramani, S. (1992). Cloning and characterization of rad21 an essential
gene of Schizosaccharomyces pombe involved in DNA double-strand-break repair.
Nucleic Acids Res. Vol.20, No.24, pp.6605-6611, ISSN:0305-1048

Borges, V., Lehane, C., Lopez-Serra, L., Flynn, H., Skehel, M., Rolef Ben-Shahar, T. &
Uhlmann, F. (2010). Hos1 deacetylates Smc3 to close the cohesin acetylation cycle.
Mol Cell. Vol.39, No.5, pp.677-688, ISSN:1097-4164.

Buard, J., Barthes, P., Grey, C. & de Massy, B. (2009). Distinct histone modifications define
initiation and repair of meiotic recombination in the mouse. *EMBO J.* Vol.28, No.17,
pp.2616-2624, ISSN:1460-2075

Buonomo, S.B., Clyne, R.K., Fuchs, J., Loidl, J., Uhlmann, F. & Nasmyth, K. (2000).
Disjunction of homologous chromosomes in meiosis I depends on proteolytic
cleavage of the meiotic cohesin Rec8 by separin. *Cell.* Vol.103, No.3, pp.387-398.

Calvente, A., Viera, A., Page, J., Parra, M.T., Gomez, R., Suja, J.A., Rufas, J.S. & Santos, J.L.
(2005). DNA double-strand breaks and homology search: inferences from a species
with incomplete pairing and synapsis. *J Cell Sci.* Vol.118, No.Pt 13, pp.2957-2963

Carramolino, L., Lee, B.C., Zaballos, A., Peled, A., Barthelemy, I., Shav-Tal, Y., Prieto, I.,
Carmi, P., Gothelf, Y., Gonzalez de Buitrago, G., Aracil, M., Marquez, G., Barbero,
J.L. & Zipori, D. (1997). SA-1, a nuclear protein encoded by one member of a novel
gene family: molecular cloning and detection in hemopoietic organs. *Gene.* Vol.195,
No.2, pp.151-159, ISSN:0378-1119

Ciosk, R., Zachariae, W., Michaelis, C., Shevchenko, A., Mann, M. & Nasmyth, K. (1998). An
ESP1/PDS1 complex regulates loss of sister chromatid cohesion at the metaphase
to anaphase transition in yeast. *Cell.* Vol.93, No.6, pp.1067-1076, ISSN:0092-8674

Ciosk, R., Shirayama, M., Shevchenko, A., Tanaka, T., Toth, A. & Nasmyth, K. (2000).
Cohesin's binding to chromosomes depends on a separate complex consisting of
Scc2 and Scc4 proteins. *Mol Cell.* Vol.5, No.2, pp.243-254, ISSN:1097-2765

Colaiacovo, M.P. (2006). The many facets of SC function during C. elegans meiosis.
Chromosoma. Vol.115, No.3, pp.195-211, ISSN:0009-5915

Cortes-Ledesma, F. & Aguilera, A. (2006). Double-strand breaks arising by replication
through a nick are repaired by cohesin-dependent sister-chromatid exchange.
EMBO Rep. Vol.7, No.9, pp.919-926, ISSN:1469-221X

Chan, R.C., Chan, A., Jeon, M., Wu, T.F., Pasqualone, D., Rougvie, A.E. & Meyer, B.J. (2003).
Chromosome cohesion is regulated by a clock gene paralogue TIM-1. *Nature.*
Vol.423, No.6943, pp.1002-1009,

Chien, R., Zeng, W., Kawauchi, S., Bender, M.A., Santos, R., Gregson, H.C., Schmiesing, J.A.,
Newkirk, D.A., Kong, X., Ball, A.R., Jr., Calof, A.L., Lander, A.D., Groudine, M.T. &
Yokomori, K. (2011). Cohesin mediates chromatin interactions that regulate

mammalian beta-globin expression. *J Biol Chem*. Vol.286, No.20, pp.17870-17878, ISSN:1083-351X

Davis, L., Rozalen, A.E., Moreno, S., Smith, G.R. & Martin-Castellanos, C. (2008). Rec25 and Rec27, novel linear-element components, link cohesin to meiotic DNA breakage and recombination. *Curr Biol*. Vol.18, No.11, pp.849-854, ISSN:0960-9822

de Massy, B., Rocco, V. & Nicolas, A. (1995). The nucleotide mapping of DNA double-strand breaks at the CYS3 initiation site of meiotic recombination in Saccharomyces cerevisiae. *EMBO J*. Vol.14, No.18, pp.4589-4598, ISSN:0261-4189

de Massy, B. (2003). Distribution of meiotic recombination sites. *Trends Genet*. Vol.19, No.9, pp.514-522, ISSN:0168-9525

Deardorff, M.A., Kaur, M., Yaeger, D., Rampuria, A., Korolev, S., Pie, J., Gil-Rodriguez, C., Arnedo, M., Loeys, B., Kline, A.D., Wilson, M., Lillquist, K., Siu, V., Ramos, F.J., Musio, A., Jackson, L.S., Dorsett, D. & Krantz, I.D. (2007). Mutations in cohesin complex members SMC3 and SMC1A cause a mild variant of cornelia de Lange syndrome with predominant mental retardation. *Am J Hum Genet*. Vol.80, No.3, pp.485-494, ISSN:0002-9297

Diaz-Martinez, L.A., Gimenez-Abian, J.F. & Clarke, D.J. (2007). Regulation of centromeric cohesion by sororin independently of the APC/C. *Cell Cycle*. Vol.6, No.6, pp.714-724, ISSN:1551-4005

Ding, D.Q., Sakurai, N., Katou, Y., Itoh, T., Shirahige, K., Haraguchi, T. & Hiraoka, Y. (2006). Meiotic cohesins modulate chromosome compaction during meiotic prophase in fission yeast. *J Cell Biol*. Vol.174, No.4, pp.499-508, ISSN:0021-9525

Dorsett, D. (2011). Cohesin: genomic insights into controlling gene transcription and development. *Curr Opin Genet Dev*. Vol.21, No.2, pp.199-206, ISSN:1879-0380

Eijpe, M., Heyting, C., Gross, B. & Jessberger, R. (2000). Association of mammalian SMC1 and SMC3 proteins with meiotic chromosomes and synaptonemal complexes. *J Cell Sci*. Vol.113 (Pt 4), pp.673-682

Eijpe, M., Offenberg, H., Jessberger, R., Revenkova, E. & Heyting, C. (2003). Meiotic cohesin REC8 marks the axial elements of rat synaptonemal complexes before cohesins SMC1beta and SMC3. *J Cell Biol*. Vol.160, No.5, pp.657-670

Ellermeier, C. & Smith, G.R. (2005). Cohesins are required for meiotic DNA breakage and recombination in Schizosaccharomyces pombe. *Proc Natl Acad Sci U S A*. Vol.102, No.31, pp.10952-10957, ISSN:0027-8424

Fernandez-Capetillo, O., Mahadevaiah, S.K., Celeste, A., Romanienko, P.J., Camerini-Otero, R.D., Bonner, W.M., Manova, K., Burgoyne, P. & Nussenzweig, A. (2003). H2AX is required for chromatin remodeling and inactivation of sex chromosomes in male mouse meiosis. *Dev Cell*. Vol.4, No.4, pp.497-508,

Fernández-Capetillo, O., Lee, A., Nussenzweig, M. & Nussenzweig, A. (2004). H2AX: the histone guardian of the genome. *DNA Repair (Amst)*. Vol.3, No.8-9, pp.959-967

Gandhi, R., Gillespie, P.J. & Hirano, T. (2006). Human Wapl is a cohesin-binding protein that promotes sister-chromatid resolution in mitotic prophase. *Curr Biol*. Vol.16, No.24, pp.2406-2417, ISSN:0960-9822

Garcia-Cruz, R., Brieno, M.A., Roig, I., Grossmann, M., Velilla, E., Pujol, A., Cabero, L., Pessarrodona, A., Barbero, J.L. & Garcia Caldes, M. (2010). Dynamics of cohesin proteins REC8, STAG3, SMC1 beta and SMC3 are consistent with a role in sister chromatid cohesion during meiosis in human oocytes. *Hum Reprod.* Vol.25, No.9, pp.2316-2327, ISSN:1460-2350

Gerton, J.L., DeRisi, J., Shroff, R., Lichten, M., Brown, P.O. & Petes, T.D. (2000). Global mapping of meiotic recombination hotspots and coldspots in the yeast Saccharomyces cerevisiae. *Proc Natl Acad Sci U S A.* Vol.97, No.21, pp.11383-11390, ISSN:0027-8424

Gomez, R., Valdeolmillos, A., Parra, M.T., Viera, A., Carreiro, C., Roncal, F., Rufas, J.S., Barbero, J.L. & Suja, J.A. (2007). Mammalian SGO2 appears at the inner centromere domain and redistributes depending on tension across centromeres during meiosis II and mitosis. *EMBO Rep.* Vol.8, No.2, pp.173-180,

Gordillo, M., Vega, H., Trainer, A.H., Hou, F., Sakai, N., Luque, R., Kayserili, H., Basaran, S., Skovby, F., Hennekam, R.C., Uzielli, M.L., Schnur, R.E., Manouvrier, S., Chang, S., Blair, E., Hurst, J.A., Forzano, F., Meins, M., Simola, K.O., Raas-Rothschild, A., Schultz, R.A., McDaniel, L.D., Ozono, K., Inui, K., Zou, H. & Jabs, E.W. (2008). The molecular mechanism underlying Roberts syndrome involves loss of ESCO2 acetyltransferase activity. *Hum Mol Genet.* Vol.17, No.14, pp.2172-2180, ISSN:1460-2083

Grelon, M., Vezon, D., Gendrot, G. & Pelletier, G. (2001). AtSPO11-1 is necessary for efficient meiotic recombination in plants. *EMBO J.* Vol.20, No.3, pp.589-600, ISSN:0261-4189

Grey, C., Baudat, F. & de Massy, B. (2009). Genome-wide control of the distribution of meiotic recombination. *PLoS Biol.* Vol.7, No.2, pp.e35, ISSN:1545-7885

Gruber, S., Haering, C.H. & Nasmyth, K. (2003). Chromosomal cohesin forms a ring. *Cell.* Vol.112, No.6, pp.765-777, ISSN:0092-8674

Guacci, V. (2007). Sister chromatid cohesion: the cohesin cleavage model does not ring true. *Genes Cells.* Vol.12, No.6, pp.693-708, ISSN:1356-9597

Gutierrez-Caballero, C., Herran, Y., Sanchez-Martin, M., Suja, J.A., Barbero, J.L., Llano, E. & Pendas, A.M. (2011). Identification and molecular characterization of the mammalian alpha-kleisin RAD21L. *Cell Cycle.* Vol.10, No.9, pp.1477-1487, ISSN:1551-4005

Haering, C.H., Lowe, J., Hochwagen, A. & Nasmyth, K. (2002). Molecular architecture of SMC proteins and the yeast cohesin complex. *Mol Cell.* Vol.9, No.4, pp.773-788

Hauf, S., Roitinger, E., Koch, B., Dittrich, C.M., Mechtler, K. & Peters, J.M. (2005). Dissociation of cohesin from chromosome arms and loss of arm cohesion during early mitosis depends on phosphorylation of SA2. *PLoS Biol.* Vol.3, No.3, pp.e69

Hayashi, M., Chin, G.M. & Villeneuve, A.M. (2007). C. elegans germ cells switch between distinct modes of double-strand break repair during meiotic prophase progression. *PLoS Genet.* Vol.3, No.11, pp.e191, ISSN:1553-7404

Herran, Y., Gutierrez-Caballero, C., Sanchez-Martin, M., Hernandez, T., Viera, A., Barbero, J.L., de Alava, E., de Rooij, D.G., Suja, J.A., Llano, E. & Pendas, A.M. (2011). The cohesin subunit RAD21L functions in meiotic synapsis and exhibits sexual dimorphism in fertility. *EMBO J*:1460-2075

Heyer, W.D., Ehmsen, K.T. & Liu, J. (2010). Regulation of homologous recombination in eukaryotes. *Annu Rev Genet.* Vol.44, pp.113-139, ISSN:1545-2948

Hogarth, C.A., Mitchell, D., Evanoff, R., Small, C. & Griswold, M. (2011). Identification and expression of potential regulators of the mammalian mitotic-to-meiotic transition. *Biol Reprod.* Vol.84, No.1, pp.34-42, ISSN:1529-7268

Hou, F. & Zou, H. (2005). Two human orthologues of Eco1/Ctf7 acetyltransferases are both required for proper sister-chromatid cohesion. *Mol Biol Cell*. Vol.16, No.8, pp.3908-3918, ISSN:1059-1524

Hunter, N. & Borts, R.H. (1997). Mlh1 is unique among mismatch repair proteins in its ability to promote crossing-over during meiosis. *Genes Dev*. Vol.11, No.12, pp.1573-1582

Ishiguro, K., Kim, J., Fujiyama-Nakamura, S., Kato, S. & Watanabe, Y. (2011). A new meiosis-specific cohesin complex implicated in the cohesin code for homologous pairing. *EMBO Rep*. Vol.12, No.3, pp.267-275, ISSN:1469-3178

James, R.D., Schmiesing, J.A., Peters, A.H., Yokomori, K. & Disteche, C.M. (2002). Differential association of SMC1alpha and SMC3 proteins with meiotic chromosomes in wild-type and SPO11-deficient male mice. *Chromosome Res*. Vol.10, No.7, pp.549-560, ISSN:0967-3849

Jessberger, R., Riwar, B., Baechtold, H. & Akhmedov, A.T. (1996). SMC proteins constitute two subunits of the mammalian recombination complex RC-1. *EMBO J*. Vol.15, No.15, pp.4061-4068, ISSN:0261-4189

Jin, H., Guacci, V. & Yu, H.G. (2009). Pds5 is required for homologue pairing and inhibits synapsis of sister chromatids during yeast meiosis. *J Cell Biol*. Vol.186, No.5, pp.713-725, ISSN:1540-8140

Jones, G.H. (1973). Light and electron microscope studies of chromosome pairing in relation to chiasma localisation in Stethophyma grossum (Orthoptera: Acrididae). *Chromosoma*. Vol.42, No.2, pp.145-162

Keeney, S., Giroux, C.N. & Kleckner, N. (1997). Meiosis-specific DNA double-strand breaks are catalyzed by Spo11, a member of a widely conserved protein family. *Cell*. Vol.88, No.3, pp.375-384

Khalil, A.M., Boyar, F.Z. & Driscoll, D.J. (2004). Dynamic histone modifications mark sex chromosome inactivation and reactivation during mammalian spermatogenesis. *Proc Natl Acad Sci U S A*. Vol.101, No.47, pp.16583-16587

Kim, B.J., Kang, K.M., Jung, S.Y., Choi, H.K., Seo, J.H., Chae, J.H., Cho, E.J., Youn, H.D., Qin, J. & Kim, S.T. (2008). Esco2 is a novel corepressor that associates with various chromatin modifying enzymes. *Biochem Biophys Res Commun*. Vol.372, No.2, pp.298-304, ISSN:1090-2104

Kim, K.P., Weiner, B.M., Zhang, L., Jordan, A., Dekker, J. & Kleckner, N. (2010). Sister cohesion and structural axis components mediate homolog bias of meiotic recombination. *Cell*. Vol.143, No.6, pp.924-937:1097-4172

Kitajima, T.S., Kawashima, S.A. & Watanabe, Y. (2004). The conserved kinetochore protein shugoshin protects centromeric cohesion during meiosis. *Nature*. Vol.427, No.6974, pp.510-517

Kitajima, T.S., Sakuno, T., Ishiguro, K., Iemura, S., Natsume, T., Kawashima, S.A. & Watanabe, Y. (2006). Shugoshin collaborates with protein phosphatase 2A to protect cohesin. *Nature*. Vol.441, No.7089, pp.46-52, ISSN:1476-4687

Klein, F., Mahr, P., Galova, M., Buonomo, S.B., Michaelis, C., Nairz, K. & Nasmyth, K. (1999). A central role for cohesins in sister chromatid cohesion, formation of axial elements, and recombination during yeast meiosis. *Cell*. Vol.98, No.1, pp.91-103

Krantz, I.D., McCallum, J., DeScipio, C., Kaur, M., Gillis, L.A., Yaeger, D., Jukofsky, L., Wasserman, N., Bottani, A., Morris, C.A., Nowaczyk, M.J., Toriello, H., Bamshad,

M.J., Carey, J.C., Rappaport, E., Kawauchi, S., Lander, A.D., Calof, A.L., Li, H.H., Devoto, M. & Jackson, L.G. (2004). Cornelia de Lange syndrome is caused by mutations in NIPBL, the human homolog of Drosophila melanogaster Nipped-B. *Nat Genet*. Vol.36, No.6, pp.631-635, ISSN:1061-4036

Kugou, K., Fukuda, T., Yamada, S., Ito, M., Sasanuma, H., Mori, S., Katou, Y., Itoh, T., Matsumoto, K., Shibata, T., Shirahige, K. & Ohta, K. (2009). Rec8 guides canonical Spo11 distribution along yeast meiotic chromosomes. *Mol Biol Cell*. Vol.20, No.13, pp.3064-3076, ISSN:1939-4586

Kuroda, M., Oikawa, K., Ohbayashi, T., Yoshida, K., Yamada, K., Mimura, J., Matsuda, Y., Fujii-Kuriyama, Y. & Mukai, K. (2005). A dioxin sensitive gene, mammalian WAPL, is implicated in spermatogenesis. *FEBS Lett*. Vol.579, No.1, pp.167-172, ISSN:0014-5793

Lafont, A.L., Song, J. & Rankin, S. (2010). Sororin cooperates with the acetyltransferase Eco2 to ensure DNA replication-dependent sister chromatid cohesion. *Proc Natl Acad Sci U S A*. Vol.107, No.47, pp.20364-20369, ISSN:1091-6490

Lee, J. & Hirano, T. (2011). RAD21L, a novel cohesin subunit implicated in linking homologous chromosomes in mammalian meiosis. *J Cell Biol*. Vol.192, No.2, pp.263-276, ISSN:1540-8140

Lengauer, C., Kinzler, K.W. & Vogelstein, B. (1997). DNA methylation and genetic instability in colorectal cancer cells. *Proc Natl Acad Sci U S A*. Vol.94, No.6, pp.2545-2550, ISSN:0027-8424

Lengronne, A., Katou, Y., Mori, S., Yokobayashi, S., Kelly, G.P., Itoh, T., Watanabe, Y., Shirahige, K. & Uhlmann, F. (2004). Cohesin relocation from sites of chromosomal loading to places of convergent transcription. *Nature*. Vol.430, No.6999, pp.573-578, ISSN:1476-

Lichten, M. (2001). Meiotic recombination: breaking the genome to save it. *Curr Biol*. Vol.11, No.7, pp.R253-256

Lin, W., Jin, H., Liu, X., Hampton, K. & Yu, H.G. (2011). Scc2 regulates gene expression by recruiting cohesin to the chromosome as a transcriptional activator during yeast meiosis. *Mol Biol Cell*. Vol.22, No.12, pp.1985-1996, ISSN:1939-4586

Losada, A., Hirano, M. & Hirano, T. (1998). Identification of Xenopus SMC protein complexes required for sister chromatid cohesion. *Genes Dev*. Vol.12, No.13, pp.1986-1997

Losada, A., Yokochi, T., Kobayashi, R. & Hirano, T. (2000). Identification and characterization of SA/Scc3p subunits in the Xenopus and human cohesin complexes. *J Cell Biol*. Vol.150, No.3, pp.405-416

Losada, A., Yokochi, T. & Hirano, T. (2005). Functional contribution of Pds5 to cohesin-mediated cohesion in human cells and Xenopus egg extracts. *J Cell Sci*. Vol.118, No.Pt 10, pp.2133-2141, ISSN:0021-9533

Llano, E., Gomez, R., Gutierrez-Caballero, C., Herran, Y., Sanchez-Martin, M., Vazquez-Quinones, L., Hernandez, T., de Alava, E., Cuadrado, A., Barbero, J.L., Suja, J.A. & Pendas, A.M. (2008). Shugoshin-2 is essential for the completion of meiosis but not for mitotic cell division in mice. *Genes Dev*. Vol.22, No.17, pp.2400-2413, ISSN:0890-9369

MacQueen, A.J., Colaiacovo, M.P., McDonald, K. & Villeneuve, A.M. (2002). Synapsis-dependent and -independent mechanisms stabilize homolog pairing during meiotic prophase in C. elegans. *Genes Dev.* Vol.16, No.18, pp.2428-2442, ISSN:0890-9369

Madigan, J.P., Chotkowski, H.L. & Glaser, R.L. (2002). DNA double-strand break-induced phosphorylation of Drosophila histone variant H2Av helps prevent radiation-induced apoptosis. *Nucleic Acids Res.* Vol.30, No.17, pp.3698-3705

Maeshima, K. & Laemmli, U.K. (2003). A two-step scaffolding model for mitotic chromosome assembly. *Dev Cell.* Vol.4, No.4, pp.467-480

Mahadevaiah, S.K., Turner, J.M., Baudat, F., Rogakou, E.P., de Boer, P., Blanco-Rodriguez, J., Jasin, M., Keeney, S., Bonner, W.M. & Burgoyne, P.S. (2001). Recombinational DNA double-strand breaks in mice precede synapsis. *Nat Genet.* Vol.27, No.3, pp.271-276

Meuwissen, R.L., Offenberg, H.H., Dietrich, A.J., Riesewijk, A., van Iersel, M. & Heyting, C. (1992). A coiled-coil related protein specific for synapsed regions of meiotic prophase chromosomes. *EMBO J.* Vol.11, No.13, pp.5091-5100, ISSN:0261-4189

Michaelis, C., Ciosk, R. & Nasmyth, K. (1997). Cohesins: chromosomal proteins that prevent premature separation of sister chromatids. *Cell.* Vol.91, No.1, pp.35-45,

Miyazaki, W.Y. & Orr-Weaver, T.L. (1994). Sister-chromatid cohesion in mitosis and meiosis. *Annu Rev Genet.* Vol.28, pp.167-187

Moldovan, G.L., Pfander, B. & Jentsch, S. (2006). PCNA controls establishment of sister chromatid cohesion during S phase. *Mol Cell.* Vol.23, No.5, pp.723-732, ISSN:1097-2765

Molnar, M., Bahler, J., Sipiczki, M. & Kohli, J. (1995). The rec8 gene of Schizosaccharomyces pombe is involved in linear element formation, chromosome pairing and sister-chromatid cohesion during meiosis. *Genetics.* Vol.141, No.1, pp.61-73, ISSN:0016-6731

Musio, A., Selicorni, A., Focarelli, M.L., Gervasini, C., Milani, D., Russo, S., Vezzoni, P. & Larizza, L. (2006). X-linked Cornelia de Lange syndrome owing to SMC1L1 mutations. *Nat Genet.* Vol.38, No.5, pp.528-530, ISSN:1061-4036

Nabeshima, K., Villeneuve, A.M. & Colaiacovo, M.P. (2005). Crossing over is coupled to late meiotic prophase bivalent differentiation through asymmetric disassembly of the SC. *J Cell Biol.* Vol.168, No.5, pp.683-689, ISSN:0021-9525

Nagao, K., Adachi, Y. & Yanagida, M. (2004). Separase-mediated cleavage of cohesin at interphase is required for DNA repair. *Nature.* Vol.430, No.7003, pp.1044-1048, ISSN:1476-4687

Nasmyth, K. & Haering, C.H. (2005). The structure and function of SMC and kleisin complexes. *Annu Rev Biochem.* Vol.74, pp.595-648

Neale, M.J. & Keeney, S. (2006). Clarifying the mechanics of DNA strand exchange in meiotic recombination. *Nature.* Vol.442, No.7099, pp.153-158

Newman, J.J. & Young, R.A. (2010). Connecting transcriptional control to chromosome structure and human disease. *Cold Spring Harb Symp Quant Biol.* Vol.75, pp.227-235, ISSN:1943-4456

Nishiyama, T., Ladurner, R., Schmitz, J., Kreidl, E., Schleiffer, A., Bhaskara, V., Bando, M., Shirahige, K., Hyman, A.A., Mechtler, K. & Peters, J.M. (2010). Sororin mediates sister chromatid cohesion by antagonizing Wapl. *Cell.* Vol.143, No.5, pp.737-749, ISSN:1097-4172

Notaridou, M., Quaye, L., Dafou, D., Jones, C., Song, H., Hogdall, E., Kjaer, S.K., Christensen, L., Hogdall, C., Blaakaer, J., McGuire, V., Wu, A.H., Van Den Berg, D.J., Pike, M.C., Gentry-Maharaj, A., Wozniak, E., Sher, T., Jacobs, I.J., Tyrer, J., Schildkraut, J.M., Moorman, P.G., Iversen, E.S., Jakubowska, A., Medrek, K., Lubinski, J., Ness, R.B., Moysich, K.B., Lurie, G., Wilkens, L.R., Carney, M.E., Wang-Gohrke, S., Doherty, J.A., Rossing, M.A., Beckmann, M.W., Thiel, F.C., Ekici, A.B., Chen, X., Beesley, J., Gronwald, J., Fasching, P.A., Chang-Claude, J., Goodman, M.T., Chenevix-Trench, G., Berchuck, A., Pearce, C.L., Whittemore, A.S., Menon, U., Pharoah, P.D., Gayther, S.A. & Ramus, S.J. (2011). Common alleles in candidate susceptibility genes associated with risk and development of epithelial ovarian cancer. *Int J Cancer*. Vol.128, No.9, pp.2063-2074, ISSN:1097-0215

Novak, I., Wang, H., Revenkova, E., Jessberger, R., Scherthan, H. & Hoog, C. (2008). Cohesin Smc1beta determines meiotic chromatin axis loop organization. *J Cell Biol*. Vol.180, No.1, pp.83-90, ISSN:1540-8140

Ong, C.T. & Corces, V.G. (2011). Enhancer function: new insights into the regulation of tissue-specific gene expression. *Nat Rev Genet*. Vol.12, No.4, pp.283-293, ISSN:1471-0064

Page, J., Suja, J.A., Santos, J.L. & Rufas, J.S. (1998). Squash procedure for protein immunolocalization in meiotic cells. *Chromosome Res*. Vol.6, No.8, pp.639-642,

Page, S.L. & Hawley, R.S. (2004). The genetics and molecular biology of the synaptonemal complex. *Annu Rev Cell Dev Biol*. Vol.20, pp.525-558

Page, J., de la Fuente, R., Gomez, R., Calvente, A., Viera, A., Parra, M.T., Santos, J.L., Berrios, S., Fernandez-Donoso, R., Suja, J.A. & Rufas, J.S. (2006). Sex chromosomes, synapsis, and cohesins: a complex affair. *Chromosoma*. Vol.115, No.3, pp.250-259, ISSN:0009-5915

Panizza, S., Tanaka, T., Hochwagen, A., Eisenhaber, F. & Nasmyth, K. (2000). Pds5 cooperates with cohesin in maintaining sister chromatid cohesion. *Curr Biol*. Vol.10, No.24, pp.1557-1564, ISSN:0960-9822

Parra, M.T., Viera, A., Gomez, R., Page, J., Benavente, R., Santos, J.L., Rufas, J.S. & Suja, J.A. (2004). Involvement of the cohesin Rad21 and SCP3 in monopolar attachment of sister kinetochores during mouse meiosis I. *J Cell Sci*. Vol.117, No.Pt 7, pp.1221-1234

Pasierbek, P., Jantsch, M., Melcher, M., Schleiffer, A., Schweizer, D. & Loidl, J. (2001). A Caenorhabditis elegans cohesion protein with functions in meiotic chromosome pairing and disjunction. *Genes Dev*. Vol.15, No.11, pp.1349-1360, ISSN:0890-9369

Pelttari, J., Hoja, M.R., Yuan, L., Liu, J.G., Brundell, E., Moens, P., Santucci-Darmanin, S., Jessberger, R., Barbero, J.L., Heyting, C. & Hoog, C. (2001). A meiotic chromosomal core consisting of cohesin complex proteins recruits DNA recombination proteins and promotes synapsis in the absence of an axial element in mammalian meiotic cells. *Mol Cell Biol*. Vol.21, No.16, pp.5667-5677

Peters, A.H., Plug, A.W., van Vugt, M.J. & de Boer, P. (1997). A drying-down technique for the spreading of mammalian meiocytes from the male and female germline. *Chromosome Res*. Vol.5, No.1, pp.66-68

Pezzi, N., Prieto, I., Kremer, L., Perez Jurado, L.A., Valero, C., Del Mazo, J., Martinez, A.C. & Barbero, J.L. (2000). STAG3, a novel gene encoding a protein involved in meiotic chromosome pairing and location of STAG3-related genes flanking the Williams-Beuren syndrome deletion. *Faseb J*. Vol.14, No.3, pp.581-592

Plug, A.W., Peters, A.H., Keegan, K.S., Hoekstra, M.F., de Boer, P. & Ashley, T. (1998). Changes in protein composition of meiotic nodules during mammalian meiosis. *J Cell Sci.* Vol.111 (Pt 4), pp.413-423, ISSN:0021-9533

Polakova, S., Cipak, L. & Gregan, J. (2011). RAD21L is a novel kleisin subunit of the cohesin complex. *Cell Cycle.* Vol.10, No.12, pp.1893, ISSN:1551-4005

Prieto, I., Suja, J.A., Pezzi, N., Kremer, L., Martinez, A.C., Rufas, J.S. & Barbero, J.L. (2001). Mammalian STAG3 is a cohesin specific to sister chromatid arms in meiosis I. *Nat Cell Biol.* Vol.3, No.8, pp.761-766

Prieto, I., Pezzi, N., Buesa, J.M., Kremer, L., Barthelemy, I., Carreiro, C., Roncal, F., Martinez, A., Gomez, L., Fernandez, R., Martinez, A.C. & Barbero, J.L. (2002). STAG2 and Rad21 mammalian mitotic cohesins are implicated in meiosis. *EMBO Rep.* Vol.3, No.6, pp.543-550, ISSN:1469-221X

Qiao, H., Lohmiller, L.D. & Anderson, L.K. (2011). Cohesin proteins load sequentially during prophase I in tomato primary microsporocytes. *Chromosome Res.* Vol.19, No.2, pp.193-207, ISSN:1573-6849

Rankin, S. (2005). Sororin, the cell cycle and sister chromatid cohesion. *Cell Cycle.* Vol.4, No.8, pp.1039-1042, ISSN:1551-4005

Redon, C., Pilch, D., Rogakou, E., Sedelnikova, O., Newrock, K. & Bonner, W. (2002). Histone H2A variants H2AX and H2AZ. *Curr Opin Genet Dev.* Vol.12, No.2, pp.162-169, ISSN:0959-437X

Revenkova, E., Eijpe, M., Heyting, C., Gross, B. & Jessberger, R. (2001). Novel meiosis-specific isoform of mammalian SMC1. *Mol Cell Biol.* Vol.21, No.20, pp.6984-6998

Revenkova, E., Eijpe, M., Heyting, C., Hodges, C.A., Hunt, P.A., Liebe, B., Scherthan, H. & Jessberger, R. (2004). Cohesin SMC1 beta is required for meiotic chromosome dynamics, sister chromatid cohesion and DNA recombination. *Nat Cell Biol.* Vol.6, No.6, pp.555-562, ISSN:1465-7392

Revenkova, E. & Jessberger, R. (2005). Keeping sister chromatids together: cohesins in meiosis. *Reproduction.* Vol.130, No.6, pp.783-790

Revenkova, E. & Jessberger, R. (2006). Shaping meiotic prophase chromosomes: cohesins and synaptonemal complex proteins. *Chromosoma.* Vol.115, No.3, pp.235-240, ISSN:0009-5915

Royo, H., Polikiewicz, G., Mahadevaiah, S.K., Prosser, H., Mitchell, M., Bradley, A., de Rooij, D.G., Burgoyne, P.S. & Turner, J.M. (2010). Evidence that meiotic sex chromosome inactivation is essential for male fertility. *Curr Biol.* Vol.20, No.23, pp.2117-2123, ISSN:1879-0445

Schalk, J.A., Dietrich, A.J., Vink, A.C., Offenberg, H.H., van Aalderen, M. & Heyting, C. (1998). Localization of SCP2 and SCP3 protein molecules within synaptonemal complexes of the rat. *Chromosoma.* Vol.107, No.8, pp.540-548

Schmitz, J., Watrin, E., Lenart, P., Mechtler, K. & Peters, J.M. (2007). Sororin is required for stable binding of cohesin to chromatin and for sister chromatid cohesion in interphase. *Curr Biol.* Vol.17, No.7, pp.630-636, ISSN:0960-9822

Shinohara, A. & Shinohara, M. (2004). Roles of RecA homologues Rad51 and Dmc1 during meiotic recombination. *Cytogenet Genome Res.* Vol.107, No.3-4, pp.201-207

Skibbens, R.V., Corson, L.B., Koshland, D. & Hieter, P. (1999). Ctf7p is essential for sister chromatid cohesion and links mitotic chromosome structure to the DNA replication machinery. *Genes Dev.* Vol.13, No.3, pp.307-319

Skibbens, R.V. (2009). Establishment of sister chromatid cohesion. *Curr Biol.* Vol.19, No.24, pp.R1126-1132, ISSN:1879-0445

Sonoda, E., Matsusaka, T., Morrison, C., Vagnarelli, P., Hoshi, O., Ushiki, T., Nojima, K., Fukagawa, T., Waizenegger, I.C., Peters, J.M., Earnshaw, W.C. & Takeda, S. (2001). Scc1/Rad21/Mcd1 is required for sister chromatid cohesion and kinetochore function in vertebrate cells. *Dev Cell.* Vol.1, No.6, pp.759-770

Stack, S.M. & Anderson, L.K. (2009). Electron microscopic immunogold localization of recombination-related proteins in spreads of synaptonemal complexes from tomato microsporocytes. *Methods Mol Biol.* Vol.558, pp.147-169, ISSN:1064-3745

Strom, L., Lindroos, H.B., Shirahige, K. & Sjogren, C. (2004). Postreplicative recruitment of cohesin to double-strand breaks is required for DNA repair. *Mol Cell.* Vol.16, No.6, pp.1003-1015

Strom, L. & Sjogren, C. (2005). DNA damage-induced cohesion. *Cell Cycle.* Vol.4, No.4, pp.536-539, ISSN:1551-4005

Strom, L. & Sjogren, C. (2007). Chromosome segregation and double-strand break repair - a complex connection. *Curr Opin Cell Biol.* Vol.19, No.3, pp.344-349, ISSN:0955-0674

Suja, J.A. & Barbero, J.L. (2009). Cohesin complexes and sister chromatid cohesion in mammalian meiosis. *Genome Dyn.* Vol.5, pp.94-116

Sumara, I., Vorlaufer, E., Gieffers, C., Peters, B.H. & Peters, J.M. (2000). Characterization of vertebrate cohesin complexes and their regulation in prophase. *J Cell Biol.* Vol.151, No.4, pp.749-762,

Szostak, J.W., Orr-Weaver, T.L., Rothstein, R.J. & Stahl, F.W. (1983). The double-strand-break repair model for recombination. *Cell.* Vol.33, No.1, pp.25-35, ISSN:0092-8674

Tanaka, T., Fuchs, J., Loidl, J. & Nasmyth, K. (2000). Cohesin ensures bipolar attachment of microtubules to sister centromeres and resists their precocious separation. *Nat Cell Biol.* Vol.2, No.8, pp.492-499

Tarsounas, M., Morita, T., Pearlman, R.E. & Moens, P.B. (1999). RAD51 and DMC1 form mixed complexes associated with mouse meiotic chromosome cores and synaptonemal complexes. *J Cell Biol.* Vol.147, No.2, pp.207-220, ISSN:0021-9525

Tonkin, E.T., Wang, T.J., Lisgo, S., Bamshad, M.J. & Strachan, T. (2004). NIPBL, encoding a homolog of fungal Scc2-type sister chromatid cohesion proteins and fly Nipped-B, is mutated in Cornelia de Lange syndrome. *Nat Genet.* Vol.36, No.6, pp.636-641, ISSN:1061-4036

Toth, A., Ciosk, R., Uhlmann, F., Galova, M., Schleiffer, A. & Nasmyth, K. (1999). Yeast cohesin complex requires a conserved protein, Eco1p(Ctf7), to establish cohesion between sister chromatids during DNA replication. *Genes Dev.* Vol.13, No.3, pp.320-333, ISSN:0890-9369

Uhlmann, F., Lottspeich, F. & Nasmyth, K. (1999). Sister-chromatid separation at anaphase onset is promoted by cleavage of the cohesin subunit Scc1. *Nature.* Vol.400, No.6739, pp.37-42, ISSN:0028-0836

Uhlmann, F., Wernic, D., Poupart, M.A., Koonin, E.V. & Nasmyth, K. (2000). Cleavage of cohesin by the CD clan protease separin triggers anaphase in yeast. *Cell.* Vol.103, No.3, pp.375-386, ISSN:0092-8674

Uhlmann, F. (2001). Chromosome cohesion and segregation in mitosis and meiosis. *Curr Opin Cell Biol.* Vol.13, No.6, pp.754-761

Uhlmann, F. (2011). Cohesin subunit Rad21L, the new kid on the block has new ideas. *EMBO Rep.* Vol.12, No.3, pp.183-184, ISSN:1469-3178

Unal, E., Heidinger-Pauli, J.M., Kim, W., Guacci, V., Onn, I., Gygi, S.P. & Koshland, D.E. (2008). A molecular determinant for the establishment of sister chromatid cohesion. *Science.* Vol.321, No.5888, pp.566-569, ISSN:1095-9203

Valdeolmillos, A.M., Viera, A., Page, J., Prieto, I., Santos, J.L., Parra, M.T., Heck, M.M., Martinez, A.C., Barbero, J.L., Suja, J.A. & Rufas, J.S. (2007). Sequential loading of cohesin subunits during the first meiotic prophase of grasshoppers. *PLoS Genet.* Vol.3, No.2, pp.e28

Van Den Berg, D.J. & Francke, U. (1993). Roberts syndrome: a review of 100 cases and a new rating system for severity. *Am J Med Genet.* Vol.47, No.7, pp.1104-1123, ISSN:0148-7299

Vega, H., Waisfisz, Q., Gordillo, M., Sakai, N., Yanagihara, I., Yamada, M., van Gosliga, D., Kayserili, H., Xu, C., Ozono, K., Jabs, E.W., Inui, K. & Joenje, H. (2005). Roberts syndrome is caused by mutations in ESCO2, a human homolog of yeast ECO1 that is essential for the establishment of sister chromatid cohesion. *Nat Genet.* Vol.37, No.5, pp.468-470, ISSN:1061-4036

Verni, F., Gandhi, R., Goldberg, M.L. & Gatti, M. (2000). Genetic and molecular analysis of wings apart-like (wapl), a gene controlling heterochromatin organization in Drosophila melanogaster. *Genetics.* Vol.154, No.4, pp.1693-1710, ISSN:0016-6731

Viera, A., Rufas, J.S., Martinez, I., Barbero, J.L., Ortega, S. & Suja, J.A. (2009a). CDK2 is required for proper homologous pairing, recombination and sex-body formation during male mouse meiosis. *J Cell Sci.* Vol.122, No.Pt 12, pp.2149-2159, ISSN:0021-9533

Viera, A., Santos, J.L. & Rufas, J.S. (2009b). Relationship between incomplete synapsis and chiasma localization. *Chromosoma.* Vol.118, No.3, pp.377-389

Waizenegger, I.C., Hauf, S., Meinke, A. & Peters, J.M. (2000). Two distinct pathways remove mammalian cohesin from chromosome arms in prophase and from centromeres in anaphase. *Cell.* Vol.103, No.3, pp.399-410, ISSN:0092-8674

Wallace, B.M.N. & Jones, G.H. (1978). Incomplete chromosome pairing and its relation to chiasma localisation in Stethophyma grossum spermatocytes. *Heredity.* Vol.40, pp.385-396

Wang, T.F., Kleckner, N. & Hunter, N. (1999). Functional specificity of MutL homologs in yeast: evidence for three Mlh1-based heterocomplexes with distinct roles during meiosis in recombination and mismatch correction. *Proc Natl Acad Sci U S A.* Vol.96, No.24, pp.13914-13919

Watanabe, Y. & Nurse, P. (1999). Cohesin Rec8 is required for reductional chromosome segregation at meiosis. *Nature.* Vol.400, No.6743, pp.461-464,

Watanabe, Y. (2004). Modifying sister chromatid cohesion for meiosis. *J Cell Sci.* Vol.117, No.Pt 18, pp.4017-4023

Watrin, E. & Peters, J.M. (2006). Cohesin and DNA damage repair. *Exp Cell Res.* Vol.312, No.14, pp.2687-2693, ISSN:0014-4827

Xu, H., Beasley, M., Verschoor, S., Inselman, A., Handel, M.A. & McKay, M.J. (2004). A new role for the mitotic RAD21/SCC1 cohesin in meiotic chromosome cohesion and segregation in the mouse. *EMBO Rep.* Vol.5, No.4, pp.378-384

Xu, H., Beasley, M.D., Warren, W.D., van der Horst, G.T. & McKay, M.J. (2005). Absence of mouse REC8 cohesin promotes synapsis of sister chromatids in meiosis. *Dev Cell.* Vol.8, No.6, pp.949-961, ISSN:1534-5807

Yuan, L., Liu, J.G., Zhao, J., Brundell, E., Daneholt, B. & Hoog, C. (2000). The murine SCP3 gene is required for synaptonemal complex assembly, chromosome synapsis, and male fertility. *Mol Cell.* Vol.5, No.1, pp.73-83

Zetka, M.C., Kawasaki, I., Strome, S. & Muller, F. (1999). Synapsis and chiasma formation in Caenorhabditis elegans require HIM-3, a meiotic chromosome core component that functions in chromosome segregation. *Genes Dev.* Vol.13, No.17, pp.2258-2270, ISSN:0890-9369

Zhang, L., Tao, J., Wang, S., Chong, K. & Wang, T. (2006). The rice OsRad21-4, an orthologue of yeast Rec8 protein, is required for efficient meiosis. *Plant Mol Biol.* Vol.60, No.4, pp.533-554, ISSN:0167-4412

Zhang, B., Jain, S., Song, H., Fu, M., Heuckeroth, R.O., Erlich, J.M., Jay, P.Y. & Milbrandt, J. (2007). Mice lacking sister chromatid cohesion protein PDS5B exhibit developmental abnormalities reminiscent of Cornelia de Lange syndrome. *Development.* Vol.134, No.17, pp.3191-3201, ISSN:0950-1991

Zhang, N., Kuznetsov, S.G., Sharan, S.K., Li, K., Rao, P.H. & Pati, D. (2008a). A handcuff model for the cohesin complex. *J Cell Biol.* Vol.183, No.6, pp.1019-1031, ISSN:1540-8140

Zhang, J., Shi, X., Li, Y., Kim, B.J., Jia, J., Huang, Z., Yang, T., Fu, X., Jung, S.Y., Wang, Y., Zhang, P., Kim, S.T., Pan, X. & Qin, J. (2008b). Acetylation of Smc3 by Eco1 is required for S phase sister chromatid cohesion in both human and yeast. *Mol Cell.* Vol.31, No.1, pp.143-151, ISSN:1097-4164

Zhang, J., Hakansson, H., Kuroda, M. & Yuan, L. (2008c). Wapl localization on the synaptonemal complex, a meiosis-specific proteinaceous structure that binds homologous chromosomes, in the female mouse. *Reprod Domest Anim.* Vol.43, No.1, pp.124-126, ISSN:0936-6768

5

Sex Chromosomes and Meiosis in Spiders: A Review

Douglas Araujo[1], Marielle Cristina Schneider[2],
Emygdio Paula-Neto[3] and Doralice Maria Cella[3]
[1]Universidade Estadual de Mato Grosso do Sul-UEMS,
Unidade Universitária de Ivinhema,
[2]Universidade Federal de São Paulo-UNIFESP, Campus Diadema,
[3]Universidade Estadual Paulista-UNESP, Campus Rio Claro,
Brazil

1. Introduction

According to Platnick (2011), the order Araneae possesses 110 families, 3,849 genera, and 42,473 species. It is divided into two suborders: Mesothelae, consisting of only one family (Liphistiidae), and Opisthothelae. The latter suborder is divided into two infra-orders: Mygalomorphae, consisting of spiders with paraxial chelicerae, and Araneomorphae, consisting of spiders with diaxial chelicerae. The latter infra-order is divided into the basal clades (Hypochilidae and Austrochiloidea), Haplogynae, and Entelegynae, which includes the majority of extant spiders (Coddington & Levi, 1991) (Fig. 1).

In the first cytogenetic studies in spiders performed by Carnoy (1885), gonads of male or female individuals were imbedded in paraffin, sectioned, and stained with Heidenhain's iron haematoxylin. Chromosome visualisation and interpretation of cytogenetic analyses were difficult to achieve with this method. Decades later, Sharma et al. (1959) and Beçak & Beçak (1960) were the first researchers to obtain spider chromosomes by the aceto-orcein or aceto-carmin squash methods.

Pinter & Walters (1971) introduced the use of colchicine solution for cytological preparations of spider testes and ovaries. This solution promotes an increase in the number of cells in mitotic and/or meiotic metaphase, the stage in which chromosomes are most easily visualised and identified. In the same decade, Brum-Zorrila & Cazenave (1974) applied 3:1 methanol:acetic acid as a fixative solution and Giemsa solution as a stain.

Matsumoto (1977) pioneered the observation of chromosomes in spider embryos. Embryos are a valuable source of mitotic metaphase cells due to the high rate of cellular division that occurs during embryonic development. There are a number of tissues that can be used in cytogenetic studies of spiders, such as gonads (testes and ovaries), cerebral ganglion, and cultured blood cells, as described by Wang & Yan (2001).

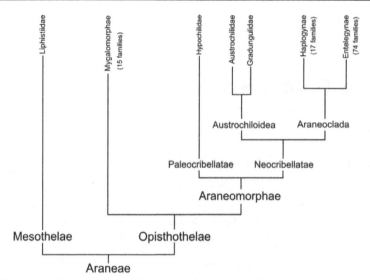

Fig. 1. Phylogenetic relationships within major clades of spiders according to Coddington & Levi (1991). The number of families was determined according to Platnick (2011).

The gonads, especially the testes, have been found to be more suitable than other tissues for karyotype analysis in the vast majority of cytogenetic investigations. Analysis of gonads allows both mitotic and meiotic chromosomes to be studied. In addition to data regarding the diploid number, length, and morphology of chromosomes, analyses of chromosomes during meiosis have contributed to the identification of types of sex chromosome systems (SCS) in spiders. This is very important in the case of Araneae due to the diversity of simple or multiple SCS that have been recorded in representatives of this order. Furthermore, in investigations of mitotic cells, it is not possible to recognise certain types of SCS, such as X0 and X_1X_2Y, by only taking into account the difference in the diploid number observed in male and female individuals. However, analysis of meiotic cells allows investigation of the behaviour of chromosomes in relation to association, synapsis, recombination, and segregation. These features are indispensable for understanding the origin and evolution of sex chromosomes.

2. Sex chromosome systems (SCS) in spiders

Currently, there are 678 cytogenetic records in spiders (www.arthropodacytogenetics.bio.br/spiderdatabase). Of these, 456 species (67.3%) have an SCS of the X_1X_20 type; 105 species (15.5%) have an X0 system; 59 species (8.7%) have an $X_1X_2X_30$ system; 10 species (1.5%) have an SCS of the X_1X_2Y type; 5 species (0.7%) have an $X_1X_2X_3X_40$ system; 5 species (0.7%) have an XY SCS; 5 species (0.7%) have an SCS of the $X_1X_2X_3Y$ type; 1 species (0.1%) has an SCS of the $X_1X_2X_3X_4X_5Y$ type; and 1 species exhibits variations of a multiple X_nY_n SCS. In 31 species (4.6%), the SCS has not been identified (Fig. 2). The number of cytogenetic records (678) in spiders is slightly higher than the number of spider species analysed chromosomally (665) because more than one type of SCS has been registered for some species.

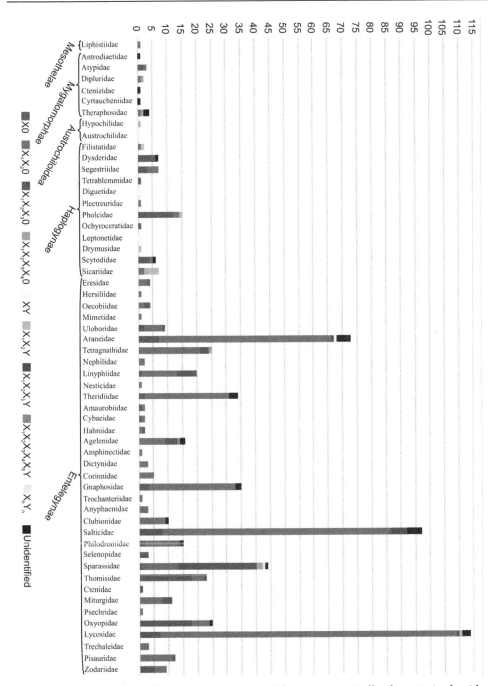

Fig. 2. Distribution of sex chromosome systems within cytogenetically characterised spider families.

2.1 Meiotic behaviour of sex chromosomes

The pioneering works describing the X_1X_20 and $X0$ SCS in spiders were those of Wallace (1900, 1905) and Berry (1906), respectively. These researchers identified the sex chromosomes based on the positive heteropycnotic behaviour of these elements during meiosis (Fig. 3). According to White (1940), the term heteropycnosis was introduced to describe the different levels of condensation and staining that certain chromosomes exhibit in the course of mitosis and/or meiosis. This heteropycnotic pattern can be positive or negative, and it is related to a high or low degree of chromosome condensation, respectively.

Manifestation of heteropycnosis is commonly visualised in the sex chromosomes, especially in male meiotic cells; the high level of chromosome condensation in these cells seems to prevent recombination between nonhomologous regions of heteromorphic sex chromosomes (McKee & Handel, 1993). However, the autosomes and female sex chromosomes can also exhibit chromatin differentiation in some stages of the cell cycle.

In spider spermatogenesis, a heteropycnotic pattern of the sex chromosomes has been recorded for roughly 25% of the species that have been cytogenetically examined, which belong to different suborders (Mygalomorphae and Araneomorphae) and families. Regardless of the type of SCS, 95% of these spider species showed positively heteropycnotic sex chromosomes in premeiotic interphase and prophase I nuclei (Fig. 3a-c, e-g, i-m) and, occasionally, also in metaphase II cells (Fig. 3d). In late meiotic stages, the sex chromosomes usually appeared to be isopycnotic (Fig. 3o-p). The heteropycnotic pattern of sex chromosomes can be used as an additional criterion to determine the type of SCS in spiders, as the number of positive heteropycnotic corpuscles frequently corresponds to the number of sex chromosomes.

It is of note that in some SCS that were originated through relatively recent rearrangements between autosomes and sex chromosomes, such as the $X_1X_2X_3Y$ system of *Evarcha hoyi* (under *Pellenes hoyi*) (Maddison, 1982), the X_nY_n system observed in *Delena cancerides* (Rowell, 1985), and the $X_1X_2X_3X_4X_5Y$ system found in *Malthonica ferruginea* (Král, 2007), only the ancient sex chromosomes exhibit positive heteropycnosis in the course of male meiosis.

Recently, positive heteropycnotic behaviour of one autosomal bivalent was verified during male meiosis of some spiders belonging to the families Dipluridae, Theraphosidae (Mygalomorphae), Diguetidae, and Sicariidae (basal Araneomorphae). According to Král et al. (2006, 2011), this bivalent could represent sex chromosomes in an early stage of differentiation because in addition to positive heteropycnosis, this bivalent showed a recurrent association with the sex chromosomes during the initial prophase I substages. Conclusive proof of the relationship between this homomorphic chromosome pair and sex determination has been obtained through ultrastructural chromosome analysis of the synaptonemal complex.

The modes of sex chromosome association and segregation are interesting features to investigate in meiotic cells. In spiders with an XY SCS, the sex chromosomes can present associations that vary from chiasmatic, such as those found in *Leptoneta infuscata* (Leptonetidae), to terminal pairing, such as in *Diguetia albolineata* and *Diguetia canities* (Diguetidae) (Král et al. 2006). An achiasmatic terminal association of the sex chromosomes has been recorded in some basal Araneomorphae with X_1X_2Y SCS. In representatives of the families Drymusidae, Filistatidae, Hypochilidae, Pholcidae, and Sicariidae, in which all sex chromosomes are biarmed, the arms of the metacentric X chromosomes showed end-to-end pairing (Fig. 3n) with arms of the tiny Y chromosome (Oliveira et al. 1997, Silva et al. 2002, Král et al. 2006). In contrast, in *Pholcus phalangioides* (Pholcidae), which carries a submetacentric X_2 chromosome, only one arm of this sex chromosome was terminally paired with the Y chromosome (Král et al. 2006).

In derivative araneomorphs that possess other types of multiple SCS, such as $X_1X_2X_3Y$ and $X_1X_2X_3X_4X_5Y$, in which the sex chromosomes do not show a high degree of morphological and/or structural differentiation due to their recent origin, the association of the sex chromosomes during meiosis can be chiasmatic between some elements and achiasmatic between others. In the $X_1X_2X_3Y$ SCS of *E. hoyi*, an interstitial or terminal chiasma was present between one arm of the large submetacentric Y chromosome and the long arm of the acrocentric X_2 chromosome (Maddison, 1982). In *M. ferruginea* (Agelenidae), the X_1, X_2, and X_3 sex chromosomes appeared as univalents that were terminally associated with the X_4X_5Y trivalent; this trivalent was composed of one large-sized Y chromosome and two small-sized acrocentric X chromosomes. Although Král (2007) did not register the presence of chiasma between these sex chromosomes due to the precocious dissociation of these elements during prophase I, ultrastructural analysis revealed the occurrence of a recombination nodule in the X_4X_5Y trivalent.

From the zygotene stage to metaphase I, the X chromosomes of multiple X SCS usually present a parallel disposition (Fig. 3 e-g, j-l), without evidence of chiasmata, and proximity between each X chromosome commonly involves the centromere region (Král et al. 2011). Occasionally, in some mygalomorph species of the family Theraphosidae, the sex chromosomes of the $X_1X_2X_3X_40$ system exhibit an end-to-end association (X_1X_2Y-like pairing), and in representatives of the family Dipluridae, the sex chromosomes can appear as univalent elements that are highly condensed and separated. In addition, an X_1X_2Y-like pairing of the sex chromosomes has been observed in *Stegodyphus lineatus* (Eresidae) and *Pax islamita* (Zodariidae), which are carriers of the $X_1X_2X_30$ and X_1X_20 SCS, respectively (Král et al. 2011).

Contrary to data revealed by conventional chromosome analyses, in which the sex chromosomes of multiple X systems appeared to exhibit a simple behaviour that involved only parallel pairing, more recent cytogenetic ultrastructural studies have supplied surprising information. Benavente and Wettstein (1980) were the first to describe the presence of junctional lamina between the X_1 and X_2 sex chromosomes of *Schizocosa malitiosa* (under *Lycosa malitiosa*); this junctional lamina was structurally similar to the synaptonemal complex, formed in the early substages of prophase I, and persisted to the late substages.

Wise (1983) also encountered evidence of a junctional lamina in prophase I in two carriers of the X_1X_20 system, *Allocosa georgicola* (under *Lycosa georgicola*) and *Rabidosa rabida* (under *Lycosa rabida*). Furthermore, a terminal association between sex chromosomes and one homomorphic bivalent was observed in *S. malitiosa* and *A. georgicola*. However, the association between sex chromosomes and a homomorphic bivalent was only explained in recent ultrastructural analyses of *Pardosa morosa* (Lycosidae) and the agelenids *Malthonica campestris* and *Malthonica silvestris* performed by Král (2007) and Král et al. (2011). These studies indicated that this homomorphic bivalent is an element that belongs to the SCS; that is, the system included one pair of homomorphic sex chromosomes in addition to the morphologically differentiated X_1 and X_2 or X_1, X_2 and X_3 sex chromosomes (Fig. 3h).

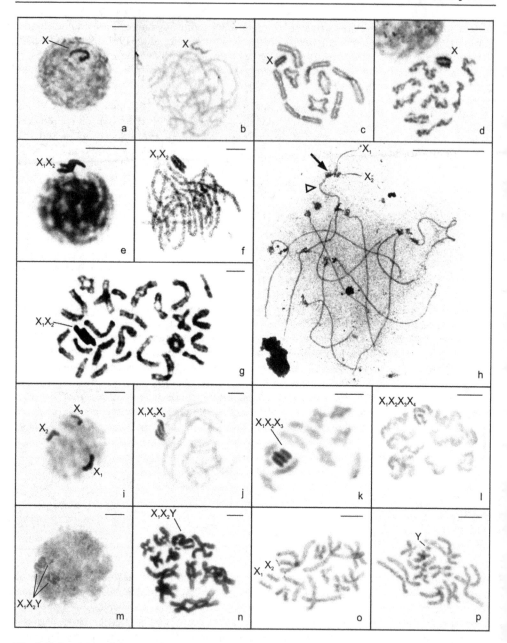

Fig. 3. Behaviour of the sex chromosomes during meiosis in male spiders. a – d. *Hogna sternalis* (X0); e. *Lycosa erythrognatha* (X_1X_20); f. *Falconina* sp. (X_1X_20); g. *Polybetes* sp. (X_1X_20); h. *Phoneutria* sp. (X_1X_20); i – k. *Trachelas* sp. ($X_1X_2X_30$); l. *Xeropigo* sp. ($X_1X_2X_3X_40$); m – p. *Loxosceles variegata* (X_1X_2Y). a, e, i, m. Premeiotic interphase nuclei. b, f, h, j. Pachytene cells. c, g, k, l, n. Diplotene cells. d, o, p. Metaphase II nuclei. In almost all cells, the sex

chromosomes can be easily recognised by their high degree of condensation and positive heteropycnosis (a-g, i-m) and/or association behaviour (e-h, j-l, n). Note the parallel pairing of the X chromosomes of the multiple sex chromosome system (e-f, j-l) and the end-to-end paring of the X_1X_2Y chromosomes (n). In h, observe the probable junctional lamina (arrow) between the X_1 and X_2 chromosomes and the terminal association of these sex chromosomes with a homomorphic bivalent (arrowhead). Scale bar=10 μm.

Although employing cytogenetic techniques to identify specific chromosomal regions has provided relevant data on spider chromosomes, only approximately 50 and 30 species were characterised with respect to the constitutive heterochromatin and nucleolar organiser regions (NORs), respectively. In these species, constitutive heterochromatin exhibited a similar distribution among autosomes and sex chromosomes, occurring mainly in pericentromeric regions. However, in at least three representatives of the basal araneomorphs with an X_1X_2Y SCS, *Pholcus phalangioides, Loxosceles intermedia,* and *Loxosceles laeta,* the Y chromosome was completely heterochromatic (Silva et al., 2002, Král et al. 2006). This reinforces the results obtained in analyses of the association behaviour of sex chromosomes during meiosis I; that is, the Y chromosome exhibits a high degree of differentiation and does not share homology with the X_1 and X_2 chromosomes. In mygalomorphs, basal and derived araneomorphs, the NORs are located predominantly on the terminal regions of one to three autosome pairs. Nevertheless, among basal araneomorph species with an X0 SCS, NORs can occur on autosomes and the X chromosome (Dysderidae, Pholcidae and Tetrablemmidae) or only on the X chromosome (Ochyroceratidae, Leptonetidae, and Scytodidae). Recently, Král et al. (2011) described the presence of NORs on the X_2 sex chromosome of derived araneomorph species belonging to the family Tetragnathidae. According to these authors, this unusual NOR localisation may be due to the translocation of rDNA cistrons from autosomes to sex chromosomes.

2.2 Early descriptions and discussions of nomenclature, function and origin

Carnoy (1885) presented the first, although inaccurate, chromosome numbers of some spider species. However, this study did not mention the existence of chromosomes that could be related to sex determination. Wallace (1900, 1909) was the first to describe double "accessory chromosomes" in a spider, *Agelenopsis naevia* (under *Agalena naevia*), and to associate the presence of these elements with sex determination. However, Wallace (1900, 1909) was not able to observe such chromosomes in females, thus concluding that male embryos were produced by fusion between a spermatozoon with accessory chromosomes and an egg without such elements, and female embryos were formed by fusion between a spermatozoon and an egg that both lack accessory chromosomes.

According to Wallace (1909), the accessory chromosomes corresponded to those that Wagner (1896) described as a "nucleolus". Subsequently, these chromosomes in spiders were also referred to as "heterochromosomes" (Montgomery, 1905) or "odd-chromosomes" (Berry, 1906). The nomenclature of "accessory chromosomes" was adopted by Painter (1914) and others. In a brief communication describing the X element in *Amaurobius* sp., King (1925) was the first to use the term "sex chromosome" in spiders. The designation of "accessory chromosomes" was still used by Hard (1939), but by the late 1940's, the nomenclature of "sex chromosomes" was definitively adopted in spiders.

An early explanation of the origin and evolution of sex chromosomes in spiders and other groups of organisms was elaborated by Montgomery (1905). According to this author, there were two types of so-called "heterochromosomes"; those that occurred in pairs in spermatogonia and then associated to form bivalents in spermatocytes (paired type), and those that were unpaired, or single, in spermatogonia and continued to be unpaired in the spermatocytes during the course of meiosis. According to this study, unpaired heterochromosomes could originate from those of the paired type through subsequent modifications. Furthermore, Montgomery (1905) stated that heterochromosomes of the paired type were derived from "ordinary chromosomes", but these heterochromosomes no longer carried out the same activities as the "ordinary chromosomes" and had the tendency to disappear. Thus, excessively minute heterochromosomes could represent the last stage before total deletion of these elements, instead of the first stage in the origin of this type of chromosome. Montgomery (1905) also suggested that heterochromosomes arose from "ordinary chromosomes" concomitantly with a change in chromosomal number, most likely from a higher to a lower number. However, the relationship between heterochromosomes and sex determination was only a hypothesis at that time because there was no record of such chromosomes in oocytes.

According to Painter (1914), the fact that accessory chromosomes were consistent with respect to their number, form, and behaviour in spider species belonging to 13 families suggested that these elements must have a very important and constant function in the life cycle of these spiders, in contrast to the autosomes, which were numerically variable among different species. Painter (1914) also noted that the accessory chromosomes were related to sex determination. The point of view of Painter (1914) was contrary to the conclusion presented by Wallace (1909), i.e., male embryos were produced by the fusion of one spermatozoon without accessory chromosomes and one egg with such elements, whereas female embryos were formed by the fusion of a spermatozoon and egg that both carried accessory chromosomes. The conclusion of Painter (1914) can be considered correct for all spider species without a Y chromosome.

2.3 Origin of the X_1X_20 sex chromosome system

The X_1X_20 SCS has been considered a plesiomorphic feature in spiders because it occurs in representatives of the phylogenetically basal family Liphistiidae (Mesothelae) (Suzuki, 1954). Various hypotheses concerning the origin of this system in spiders have been put forth.

Revell (1947) was the first to suggest that the X_1X_20 SCS most likely originated from an X0 system in spiders, considering the proposition of White (1940), who suggested that duplication of the X chromosome from an X0 system gave rise to the multiple X chromosome systems (Fig. 4). This hypothesis was based on the similarity of the sizes of X chromosomes and the probable homology between the X chromosomes during prophase I in a multiple sex chromosome system. However, Revell (1947) verified the presence of multiple X chromosomes of different sizes in *Tegenaria*, and he suggested that these X chromosomes had undergone evolutionary differentiation after originating from an X0 system (Fig. 5).

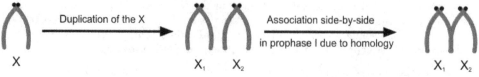

Fig. 4. Interpretive scheme of the origin of the X_1X_20 SCS based on descriptions of White (1940).

Fig. 5. Schematic representation of the origin of the X_1X_20 SCS based on descriptions of Revell (1947).

Studying the meiotic cells of *Araneus quadratus* (under *Aranea reaumuri*), Pätau (1948) suggested that there was no indication of homology between the X_1 and X_2 chromosomes; however, this author did not exclude the possibility of partial homology. Absence of homology between the X_1 and X_2 chromosomes was corroborated by Hackman (1948), Suzuki (1952) and Mittal (1964). Moreover, Pätau (1948) proposed that the X_1X_20 SCS was formed by centric fission of a large X chromosome in an X0 system.

Due to the fact that all X chromosomes of X0 systems that were registered at that time exhibited subterminal or terminal centromeres, Pätau (1948) suggested that the smaller X_1 and X_2 chromosomes had originated from the X0 system not only by simple centric fission, but through additional rearrangements such as 1) centric fragmentation and fission in the long arm terminal region followed by inversion of the long chromosome segment, resulting in a dicentric chromosome; 2) fission in the middle region of the dicentric chromosome, forming two acrocentric X_1 and X_2 chromosomes of similar size (Fig. 6).

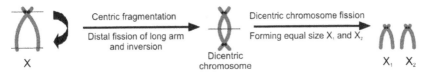

Fig. 6. Interpretive scheme of the origin of the X_1X_20 SCS based on descriptions of Pätau (1948).

Bole-Gowda (1950) asserted that the X_1X_20 SCS originated from an X0 system in the ancestor of spiders by fission in the middle of the X chromosome producing an acentric chromosome segment that then translocated to a supernumerary centric fragment. These rearrangements produced X_1 and X_2 chromosomes of similar sizes, one of which retained the original centromere of the X chromosome, while the other kept the supernumerary centromere (Fig. 7). Suzuki (1954) was in agreement with the hypothesis described by Pätau (1948) but disagreed with the assertion of Bole-Gowda (1950) regarding the origin of the X_1X_20 system in spiders because no explanation was provided for the origin of the supernumerary centric fragment involved in this hypothesis.

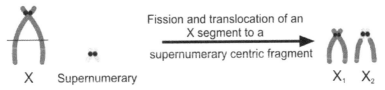

Fig. 7. Schematic representation of the origin of the X_1X_20 SCS based on the descriptions of Bole-Gowda (1950).

Considering the terminal position of the centromere and the size of the X_1 and X_2 chromosomes, Postiglioni & Brum-Zorrilla (1981) suggested that non-disjunction or duplication of a single telocentric X chromosome may have been responsible for the origin of the X_1X_20 SCS of *Lycosa* sp.3 (Fig. 8a). According to these researchers, the lack of homology between these chromosomes was most likely due to the occurrence of other rearrangements. However, the authors did not exclude the possibility that the origin of the X_1X_20 system of *Lycosa* sp.3. could have occurred by centric fission of a metacentric X chromosome (Fig. 8b).

Although previous authors reported that female X chromosomes form normal bivalents during meiosis (Hackman, 1948; Pätau, 1948; Sharma et al., 1959), Král (2007) and Král et al. (2011) found that the female X chromosomes of entelegyne and mygalomorph spiders paired during meiosis and also exhibited heterochromatinisation similar to that observed in male meiosis. This event could prevent recombination between the X chromosomes and accelerate their differentiation, a process that is consistent with the hypothesis of Postiglioni & Brum-Zorrilla (1981).

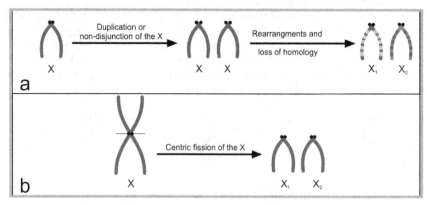

Fig. 8. Interpretive scheme of the origin of the X_1X_20 SCS based on descriptions of Postiglioni & Brum-Zorrilla (1981).

In addition to these hypotheses, some researchers have considered the X_1X_20 SCS to have originated secondarily from other multiple SCS. Oliveira et al. (2007) proposed that the X_1X_20 system could have arisen by gradual heterochromatinisation and erosion of the Y chromosome (Fig. 9). This hypothesis was based on the fact that the pholcid *Spermophora senoculata*, which exhibits an X_1X_2Y system, was considered phylogenetically basal (Bruvo-Mararic et al., 2005) in relation to *Crossopriza lyoni*, which shows an X_1X_20 SCS.

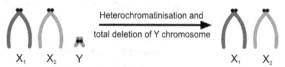

Fig. 9. Schematic representation of the origin of the X_1X_20 SCS based on the descriptions of Oliveira et al. (2007).

In some species of the genus *Malthonica* (Agelenidae), Král (2007) encountered three types of SCS, X_1X_20, $X_1X_2X_30$, and $X_1X_2X_3X_4X_5Y$, and therefore suggested that in this genus, the $X_1X_2X_30$ condition was ancestral and gave rise to the X_1X_20 system through a tandem fusion between two X chromosomes (Fig. 10). This proposition was based on the peculiar meiotic behaviour of the X chromosomes belonging to the $X_1X_2X_30$ SCS, the great difference in sizes between the X chromosomes of the X_1X_20 SCS, and the presence of the $X_1X_2X_30$ system in *Tegenaria parietina*, which is morphologically closely related to *Malthonica*.

Fig. 10. Interpretive scheme of the origin of the X_1X_20 SCS based on descriptions presented by Král (2007).

2.4 Origin of the $X_1X_2X_30$ sex chromosome system

Revell (1947) found $2n\male=40+X_1X_20$ in *Tegenaria atrica* and $2n\male=40+X_1X_2X_30$ in *Tegenaria domestica*. Considering that the number of autosomes was constant (40 autosomes) in these two *Tegenaria* species, the author suggested that the X1X2X30 SCS originated from the X_1X_20 system, which was derived from an X0 system. However, this author did not state the chromosome rearrangements that were responsible for the origin of the $X_1X_2X_30$ system. The existence of X chromosomes of different lengths in species with an $X_1X_2X_30$ system was considered by Revell (1947) as evidence that the multiple X chromosomes of the $X_1X_2X_30$ system were modified after the origination of this SCS.

According to Pätau (1948), in the *Tegenaria* species studied by Revell (1947), the largest X chromosome found in *T. atrica* (X_1X_20) was equivalent to the sum of the lengths of the two smallest X chromosomes observed in *T. domestica* ($X_1X_2X_30$). In this case, an X_1X_20 SCS could give rise to an $X_1X_2X_30$ system through rearrangements similar to those proposed by Pätau (1948) for the origin of the X_1X_20 SCS (Fig. 11, 6). This explanation was supported by Suzuki (1954) and Sharma et al. (1959). However, these authors analysed *Selenops radiatus*, which showed three X chromosomes of equal size, and proposed that the similarity in the length of the sex chromosomes in some species with the $X_1X_2X_30$ system could be the result of additional rearrangements after their origin from an X_1X_20 system.

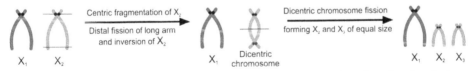

Fig. 11. Schematic representation of the origin of the $X_1X_2X_30$ SCS based on the descriptions of Pätau (1948).

Sharma et al. (1959) demonstrated that the six X chromosomes in the zygotene and pachytene stages in *Selenops radiatus* females (males were $X_1X_2X_30$) did not show any positive heteropycnosis and formed only normal bivalents but did not form multivalents; similar meiotic behaviour of chromosomes was observed in female meiotic cells of a species with four X chromosomes (males were X_1X_20) by Hackman (1948), Pätau (1948) and Suzuki (1954), reinforcing the suggestion that there was no homology between the X chromosomes of the X_1X_20 and $X_1X_2X_30$ SCS.

Bole-Gowda (1952) proposed that the $X_1X_2X_30$ system found in *Heteropoda venatoria* arose from an ancestor with an X_1X_20 system by translocation of a segment of one X chromosome of the X_1X_20 system and one supernumerary centric fragment (Fig. 12). This is similar to this author's hypothesis for the origin of the X_1X_20 system from the X0 system.

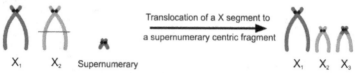

Fig. 12. Interpretive scheme of the origin of the $X_1X_2X_30$ SCS based on descriptions of Bole-Gowda (1952).

To explain the origin of the $X_1X_2X_30$ SCS found in *Lycosa* sp. (*thorelli* group), Postiglioni & Brum-Zorrilla (1981) hypothesised that non-disjunction of one X chromosome of the X_1X_20 system, followed by loss of homology between the X chromosomes had occurred (Fig. 13). To corroborate the hypothesis of non-disjunction, the authors cited the observation of a particular behaviour involving early condensation and isolation of one X chromosome at the pachytene stage. Taking into account that the X_1X_20 system could be ancestral in spiders, Postiglioni & Brum-Zorrilla (1981) suggested that each sex chromosome would undergo independent mutations during the course of evolution; however, if one of the X chromosomes had suffered a recent non-disjunction, these last two elements would present similar behaviour during meiotic prophase I, and the X chromosome not involved in the event would conserve its individuality and would appear condensed in early stages.

Fig. 13. Schematic representation of the origin of the $X_1X_2X_30$ SCS based on descriptions presented by Postiglioni & Brum-Zorrilla (1981).

Parida & Sharma (1986) observed that in some spider species with an $X_1X_2X_30$ SCS, two X chromosomes were small and one was large, suggesting that this system was derived from a small fragment of the X_1 or X_2 chromosome of an X_1X_20 system by deletion of most of its chromatin and a subsequent increase in length by duplications (Fig. 14).

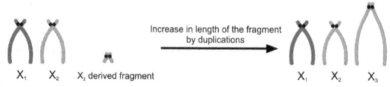

Fig. 14. Interpretive scheme of the origin of the $X_1X_2X_30$ SCS based on descriptions of Parida & Sharma (1986).

2.5 Origin of the $X_1X_2X_3X_40$ sex chromosome system

Data & Chatterjee (1983, 1988) were the first to record an $X_1X_2X_3X_40$ SCS in spiders. This SCS was found in *Metellina segmentata* (*Meta segmentata*) (Tetragnathidae) and *Bhutaniella sikkimensis* (*Heteropoda sikkimensis*) (Sparassidae). In 1988, these researchers presented a proposal for the origin of the $X_1X_2X_3X_40$ system: duplication or non-disjunction of one X chromosome of the $X_1X_2X_30$ SCS, with subsequent loss of homology (Fig. 15). This proposition was similar to that formulated by Postiglioni & Brum-Zorrilla (1981) to explain the origin of the X_1X_20 and $X_1X_2X_30$ systems.

Fig. 15. Interpretive scheme of the origin of the $X_1X_2X_3X_40$ SCS based on descriptions of Datta & Chatterjee (1988).

In *Diplura* species, which employ an $X_1X_2X_3X_40$ system, Král et al. (2011) found that one heteropycnotic and one isopycnotic portion could be distinguished in the X_1 and X_2 chromosomes, most likely corresponding to the original X_1 and X_2 chromosomes and the original autosomes, respectively, indicating that the X_1 and X_2 sex chromosomes originated by sex chromosome/autosome translocation. Additionally, Král et al. (2011), suggested that this system probably originated by duplication of the X_1X_20 system via non-disjunctions or polyploidisation (Fig. 16) based on the chromomere pattern and size of the sex chromosomes observed in the $X_1X_2X_3X_40$ *Diplura*.

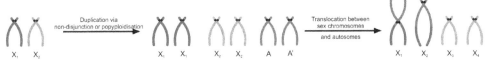

Fig. 16. Schematic representation of the origin of the $X_1X_2X_3X_40$ SCS based on the descriptions of Král et al. (2011).

2.6 Origin of the X0 sex chromosome system

Hackman (1948) found the supposed first case of a metacentric X chromosome in spiders in *Oxyopes ramosus* and noted that the X0 SCS verified in *Oxyopes* (metacentric), *Myrmarachne*, *Misumena*, and *Xysticus* (acrocentric) had most likely been derived from the X_1X_20 system in two ways:

1. The metacentric X of the X0 system could have been derived by centric fusion between X_1 and X_2 chromosomes (Fig. 17). This mechanism was also employed by several authors (Bole-Gowda, 1952; Suzuki, 1954; Postiglioni & Brum-Zorrilla, 1981; Řezáč et al., 2006; Král et al., 2011; Stávale et al. 2011) to explain the origin of the X0 SCS, which involves a metacentric X, in many spider groups.

2. The acrocentric X of the X0 system could have originated through gradual elimination of one X chromosome of the X_1X_20 SCS, as suggested by Suzuki (1952, 1954). This author put forth this proposition based on the fact that some thomisid species with an X_1X_20 system presented gradual differences between the lengths of X_1 and X_2 chromosomes (with both showing the same, slightly different or markedly different sizes). Furthermore, some species even exhibited an X0 system, suggesting that elimination of one X of the X_1X_20 system had taken place in the course of evolution (Fig. 18).

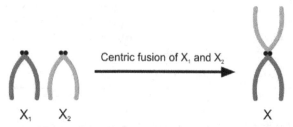

Fig. 17. Schematic representation of the origin of the X0 SCS based on descriptions presented by Hackman (1948).

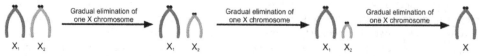

Fig. 18. Interpretive scheme of the origin of the X0 SCS based on descriptions of Hackman (1948).

Bole-Gowda (1950) proposed that the X0 system found in several spider species could have evolved by reciprocal translocation between the X_1 and X_2 chromosomes, preceded by distal fission in one sex chromosome and proximal fission in the other X, giving rise to a large acrocentric X chromosome, as found in *Oxyopes hindostanicus*; the centric fragment produced in this process was lost (Fig. 19).

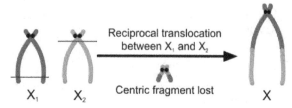

Fig. 19. Schematic representation of the origin of the X0 SCS based on descriptions presented by Bole-Gowda (1950).

Datta & Chatterjee (1989, 1992) proposed that the X0 system found in lycosid and uloborid spiders originated from the X_1X_20 SCS by centric fusion of the X_1 and X_2 chromosomes, followed by pericentric inversion (Fig. 20a) or partial deletion (Fig. 20b) in one of the X chromosome arms, giving rise to an acrocentric element. Alternatively, the acrocentric X chromosome could have originated from tandem fusion between the X_1 and X_2 chromosomes (Fig. 20c).

Tandem fusion was postulated as the mechanism involved in the derivation of the acrocentric X chromosome (X0 system) of *Zodarion* from the acrocentric X_1 and X_2 chromosomes (X_1X_20 system) present in species of the same genus (Pekár & Král, 2001). Pekár et al. (2005) proposed that two positive heteropycnotic bodies observed in the premeiotic interphase nuclei of two *Zodarion* species with the X0 SCS were segments of the X chromosome that corresponded to the original X_1 and X_2.

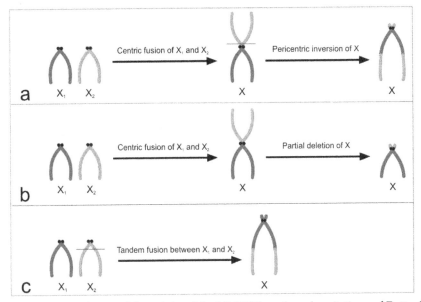

Fig. 20. Interpretive scheme of the origin of the X0 SCS based on descriptions of Datta & Chatterjee (1989, 1992).

According to Král et al. (2006), the XY SCS, which was originated from the X_1X_2Y system, gave rise to an X0 SCS through loss of the Y chromosome (Fig. 21). The SCS of the X0 type was found in many pholcids and Scytodes (Scytodidae).The evidence for this degeneration and complete elimination of the Y chromosome was the high level of constitutive heterochromatin detected in this chromosome in *Pholcus phalangioides*.

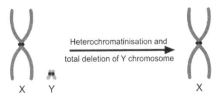

Fig. 21. Schematic representation of the origin of the X0 SCS based on descriptions of Král et al. (2006).

2.7 Origin of the XY sex chromosome system

Řezáč et al. (2006) described the first XY SCS in a spider of the genus *Atypus* (Mygalomorphae). According to these authors, this neo-XY system was formed from an X0 system, which was recorded in other species of this genus and involved rearrangements between the X chromosome and autosomes.

Král et al. (2006) studied the evolution of the chromosomes in several basal araneomorphs and formulated some hypotheses about SCS evolution. These authors discovered that SCS that include a Y chromosome are more common in spiders than previously believed, at least in basal araneomorphs. Several species with XY and X_1X_2Y SCS were described in many families, and the authors highlighted the fact that in some species previously described as carriers of the X_1X_20 system, the tiny Y chromosome could have been neglected. Based on the diversity of SCS in basal araneomorphs, specifically in pholcids, Král et al. (2006) proposed that the X_1X_2Y SCS involving metacentric chromosomes in this group of spiders was similar to that found in Filistatidae and *Loxosceles* and was converted into an XY SCS. First, by pericentric inversion transforming one of the metacentric X chromosomes into a subtelo- or acrocentric form, the X_1X_2Y became similar to that described in *Pholcus phalangioides* (Pholcidae). Subsequently, the other X chromosome of the X_1X_2Y system was also pericentrically inverted, forming a hypothetical configuration in which both X chromosomes of the X_1X_2Y system exhibit acrocentric morphology. In the next step, centric fusion between the acrocentric X chromosomes of the hypothetical X_1X_2Y SCS occurred, generating an XY system, as found in *Smeringopus pallidus* (Pholcidae) and *Diguetia* (Diguetidae). In *Diguetia albolineata*, which exhibits an XY SCS, both arms of the metacentric Y chromosome paired with only one arm of the metacentric X during meiosis. In *Diguetia canities*, the metacentric X chromosome was supposedly pericentrically inverted, resulting an acrocentric element (Fig. 22).

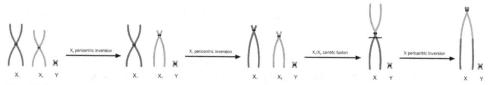

Fig. 22. Interpretive scheme of the origin of the XY SCS of Pholcidae and Diguetidae based on the descriptions of Král et al. (2006).

The proposed origin of the XY SCS observed in *Leptoneta* (Leptonetidae) was quite different. The XY system was believed to have originated from the X0 system, not from the X_1X_2Y system, in a mechanism involving translocation between the X and one autosome constituting the neo X chromosome; the homolog of the autosome involved in the rearrangement formed the Y chromosome (Fig. 23). This hypothesis put forth by Král et al. (2006) was based on the fact that the distal part of the X chromosome in *Leptoneta*, which has an XY SCS (probably corresponding to the translocated autosome), presented a different pattern of condensation during meiosis I. This last characteristic was not detected in the sex chromosomes of *Diguetia* species.

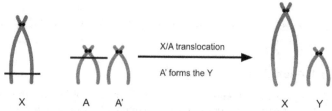

Fig. 23. Schematic representation of the origin of the XY SCS of Leptonetidae based on the descriptions of Král et al. (2006).

2.8 Origin of the X_1X_2Y sex chromosome system

Silva (1988) was the first to record an X_1X_2Y SCS in spiders. Although the author noted the possibility that the small acrocentric chromosome found in *Loxosceles laeta* was a supernumerary chromosome, she concluded that this element could correspond to the Y chromosome of the X_1X_2Y SCS. A very general citation of many types of rearrangements that could be involved in the origin of the X_1X_2Y system was presented, though without explaining the sequence of steps involved in the evolution of this SCS in spiders. Subsequently, Silva et al. (2002) proposed that the X_1X_2Y system found in *Loxosceles* (Sicariidae) was derived from an X_1X_20 SCS through a mechanism involving translocations between X chromosomes and autosomes; however, they did not provide details associated with this process.

2.9 Origin of the $X_1X_2X_3Y$ sex chromosome system

Maddison (1982) described an SCS of the $X_1X_2X_3Y$ type in five species of Salticidae and, surprisingly, verified that in *Evarcha hoyi* (under *Pellenes hoyi*), some individuals presented the $X_1X_2X_3Y$ system, whereas others showed the X_1X_20 system.

According to Maddison (1982), during the process of arising from the X_1X_20 system, the X_1 chromosome of the $X_1X_2X_3Y$ system remained unaltered, while the X_2 chromosome became tandemly fused (or centrically fused followed by pericentric inversion) with an autosome that then constituted the distal portion of the neo X_2 long arm. The homolog of this autosome involved in the autosome/X_2 fusion became centrically fused with other autosome, forming the Y chromosome (short and long arms, respectively); the homolog of this last autosome (the Y long arm) became the X_3 chromosome without undergoing modifications (Fig. 24). This hypothesis was based on the difference in diploid number detected in salticid species (i.e., individuals with the X_1X_20 system had two additional autosomal pairs when compared with individuals with the $X_1X_2X_3Y$ system). These two autosomal pairs could be the pairs involved in fusions with the original X_1X_20 system. Furthermore, the meiotic features of the sex chromosomes, such as pycnosis, achiasmatic and chiasmatic pairing, and segregation, were considered to represent additional supporting evidence of this mode of origin of the $X_1X_2X_3Y$ SCS.

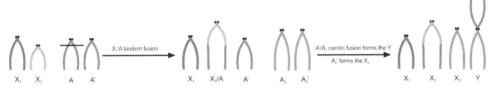

Fig. 24. Interpretive scheme of the origin of the $X_1X_2X_3Y$ SCS based on descriptions of Maddison (1982).

2.10 Origin of the $X_1X_2X_3X_4X_5Y$ sex chromosome system

The $X_1X_2X_3X_4X_5Y$ SCS verified in *Malthonica ferruginea* by Král (2007) formed a multivalent association in the pachytene stage, which was constituted by three univalents ($X_1X_2X_3$) and one trivalent (X_4X_5Y). In the trivalent, a synaptonemal complex was observed between the metacentric Y and the X_4X_5 chromosomes. Pairing between the univalents and the trivalent occurred end-to-end and involved only the X chromosomes. This mode of pairing was not visualised at the end of the diplotene stage. Observation of end-to-end pairing between the X chromosomes and a specific bivalent during meiosis of related species (Král, 2007), as well as in other spider groups (Benavente & Wettstein, 1977; Wise, 1983), compelled Král (2007) to hypothesise that the $X_1X_2X_3X_4X_5Y$ SCS originated from the $X_1X_2X_30$ system. First, the three univalents ($X_1X_2X_3$) paired with a proto-X proto-Y bivalent or homomorphic sex chromosome pair (a bivalent that did not present morphological peculiarities but probably exhibited molecular differentiation into new sex chromosomes). Subsequently, the proto-Y chromosome underwent centric fusion with an autosomal element, forming a neo Y (metacentric) that maintained the pairing with the protoX (newly denominated X_4). The homolog of the autosome element involved in centric fusion with the protoY was designated the X_5 chromosome, which paired with its homologous arm in the neo Y (Fig. 25). According to Král (2007), these proto-X and proto-Y chromosomes probably represented ancestral chromosomes involved in the origin of the multiple X chromosomes by nondisjunction.

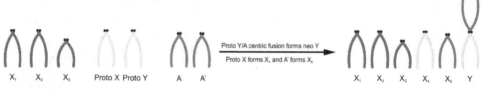

Fig. 25. Schematic representation of the origin of the $X_1X_2X_3X_4X_5Y$ SCS based on the descriptions of Král (2007).

2.11 Origin of multiple X_nY_n sex chromosome systems

Rowell (1985, 1988, 1990, 1991), Hancock & Rowell (1995), and Sharp & Rowell (2007) described a range of multiple X_nY_n SCS in populations of the social spider *Delena cancerides* (Sparassidae). Considering that some populations of this species presented an $X_1X_2X_30$ SCS, the authors provided an explanation for the origin of the multiple X and multiple Y chromosomes from the $X_1X_2X_30$ SCS, which is a system also found in other sparassid species. The first step in this process was centric fusion between two X chromosomes of the original $X_1X_2X_30$ SCS, giving rise to one metacentric X chromosome, while one X chromosome was unchanged. Subsequently, the telocentric X underwent centric fusion with one telocentric autosome, and the homolog of this autosome became part of the SCS. Subsequently, a series of centric fusions between the newly formed sex chromosome and telocentric autosomes gave rise to populations with $X_1X_2X_3Y$, $X_1X_2X_3X_4Y_1Y_2$ (Fig. 26), and multiple X and Y chromosome systems. In all of these populations, the metacentric X formed by the fusion of two original X chromosomes was a univalent, and the neoX and neoY chromosomes, which had originated through centric fusions between autosomes, formed chromosomal chains with different numbers of elements involved (from 3 to 19 elements) during meiosis.

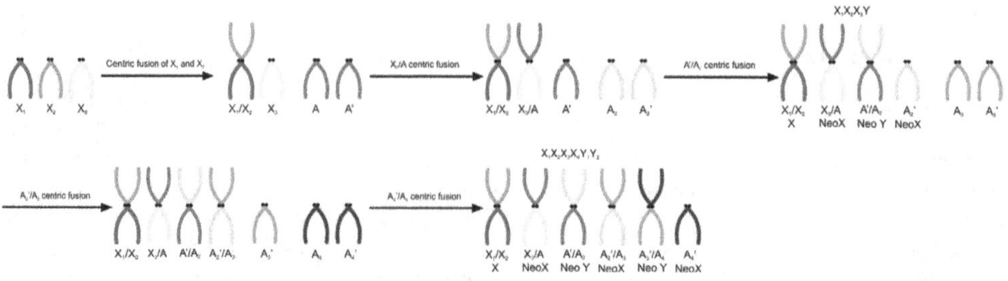

Fig. 26. Interpretive scheme of the origin of the X_nY_n SCS based on descriptions of Rowell (1985, 1988, 1990, 1991), Hancock & Rowell (1995), and Sharp & Rowell (2007).

3. Conclusion

Despite the fact that spider meiosis analyses have been carried out for more than 100 years, relatively little has been learned about sex chromosome origin and evolution in this group. The vast majority of studies on this topic have been based solely on assumptions or on basic chromosomal characteristics (chromosome length, number, meiotic condensation, and meiotic segregation). Recent advances in spider cytogenetics, such as ultrastructural analysis of cells during meiosis and examination of female meiosis, have added new insights into SCS evolution. However, only 665 (~1.6%) of the 42,423 known spider species (110 families) have been chromosomally characterised. Fifty-four families have not been studied cytogenetically, resulting in several gaps in the existing hypotheses on sex chromosome evolution.

To provide a broader knowledge base leading to better inferences regarding the origin and evolution of sex chromosomes, further efforts involving spider meiosis analysis should include a broader range of species and use conventional, ultrastructural (synaptonemal complex), and molecular (rDNA and telomere FISH and chromosome painting) cytogenetic techniques.

4. Acknowledgments

Financial support was provided by the Conselho Nacional de Desenvolvimento Científico e Tecnológico (CNPq - 471821/2008-0) and the Fundação de Amparo à Pesquisa do Estado de São Paulo (FAPESP – 2008/055633-0 and 2010/14193-7).

5. References

Araujo, D.; Schneider, M.C.; Paula-Neto, E. & Cella, D.M. (October 2011). The spider cytogenetic database. In: *Arthropoda cytogenetics website*, 07.10.2011, Available from www.arthropodacytogenetics.bio.br/spiderdatabase

Carnoy, J.B. (1885). La cytodiérèse chez les arthropodes. *La Cellule*, Vol.1, pp. 191-440, ISSN 0008-8757

Beçak, W. & Beçak, M.L. (1960). Constituição cromossômica de duas espécies de aranhas do gênero "Loxosceles". *Revista Brasileira de Biologia*, Vol.20, No.4 (December 1960), pp. 425-427, ISSN 0034-7108

Benavente, R. & Wettstein, R. (1977). An ultrastructural cytogenetic study on the evolution of sex chromosomes during the spermatogenesis of *Lycosa malitiosa* (Arachnida). *Chromosoma*, Vol.64, No.3, (September 1977), pp. 255-277, ISSN 0009-5915

Benavente, R. & Wettstein, R. (1980). Ultrastructural characterization of the sex chromosomes during spermatogenesis of spider having holocentric chromosomes and a long diffuse stage. *Chromosoma*, Vol.77, No.1, (February 1980), pp. 69-81, ISSN 0009-5915

Berry, E.H. (1906). The "acessory chromosome" in *Epeira*. *Biological Bulletin*, Vol.11, No.4, (September 1906), pp. 193-201, ISSN 0006-3185

Bole-Gowda, B.N. (1950). The chromosome study in the spermatogenesis of two lynx-spiders (Oxyopidae). *Proceedings of the Zoological Society of Bengal*, Vol.3, No.2, (September 1950), pp. 95-107, ISSN 0373-5893

Bole-Gowda, B.N. (1952). Studies on the chromosomes and the sex-determining mechanism in four hunting spiders (Sparassidae). *Proceedings of the Zoological Society of Bengal*, Vol.5, No.1, (March 1952), pp. 51-70, ISSN 0373-5893

Brum-Zorrilla, N. & Cazenave, A.M. (1974). Heterochromatin localization in the chromosomes of *Lycosa malitiosa* (Arachnida). *Experientia*, Vol.30, No.1, (January 1974) pp. 94-95, ISSN 0014-4754

Bruvo-Mađarić, B.; Huber, B.A.; Steinacher, A. & Pass, G. Phylogeny of pholcid spiders (Araneae: Pholcidae): Combined analysis using morphology and molecules. *Molecular Phylogenetics and Evolution*, Vol.37, No.3, (December 2005), pp. 661-673, ISSN 1055-7903

Coddington, J.A. & Levi, H.W. (1991). Systematics and evolution of spiders (Araneae). *Annual Review of Ecology and Systematics*, Vol.22, (November 1991), pp. 565-592, ISSN 0066-4162

Datta, S.N. & Chatterjee, K. (1983). Chromosome number and sex-determining system in fifty-two species of spiders from North-East India. *Chromosome Information Service*, Vol.35, pp. 6-8, ISSN 0574-9549

Datta, S.N. & Chatterjee, K. (1988) Chromosomes and sex determination in 13 araneid spiders of North-Eastern India. *Genetica*, Vol.76, No.2, (March 1988), pp. 91-99, ISSN 0016-6707

Datta, S.N. & Chatterjee, K. (1989). Study of meiotic chromosomes of four hunting spiders of north eastern India. *Perspectives in Cytology and Genetics*, Vol.6, pp. 414-424, ISSN 0970-4507

Datta, S.N. & Chatterjee, K. (1992). Chromosomes and sex determination in three species of spinner spiders from northeastern India. *Cell and Chromosome Research*, Vol.15, No.2, pp. 64-69

Hackman, W. (1948). Chromosomenstudien an Araneen mit besonderer berücksichtigung der geschlechtschromosomen. *Acta Zoologica Fennica*, Vol.54, pp. 1-101, ISSN 0001-7299

Hancock, A.J. & Rowell, D.M. (1995). A chromosomal hybrid zone in the Australian huntsman spider, *Delena cancerides* (Araneae: Sparassidae). Evidence for a hybrid zone near Canberra, Australia. *Australian Journal of Zoology*, Vol.43, No.2, pp. 173-180, ISSN 0004-959X

Hard, W.L. (1939). The spermatogenesis of the lycosid spider *Schizocosa crassipes* (Walckenaer). *Journal of Morphology*, Vol.65, No.1, (July 1939), pp. 121-153, ISSN 0362-2525

King, S.D. (1925). Spermatogenesis in a spider (*Amaurobius* sp.). *Nature*, Vol.116, (October 1925), pp. 574-575, ISSN 0028-0836

Král, J.; Musilová, J.; Šťálavský, F.; Řezáč, M.; Akan, Z.; Edwards, R.L.; Coyle, F.A. & Almerje, C.R. (2006). Evolution of the karyotype and sex chromosome systems in basal clades of araneomorph spiders (Araneae: Araneomorphae). *Chromosome Research*, Vol.14, No.8, (December 2006), pp. 859-880, ISSN 0967-3849

Král, J. (2007). Evolution of multiple sex chromosomes in the spider genus *Malthonica* (Araneae: Agelenidae) indicates unique structure of the spider sex chromosome systems. *Chromosome Research*, Vol.15, No.7 (November 2007), pp. 863-879, ISSN 0967-3849

Král, J.; Kořínková, T.; Forman, M. & Krkavcová, L. (2011). Insights into the meiotic behavior and evolution of multiple sex chromosome systems in spiders. *Cytogenetic and Genome Research*, Vol.133, No.1, (March 2011), pp. 43-66, ISSN 1424-8581

Maddison, W.P. (1982). XXXY sex chromosomes in males of the jumping spider genus *Pellenes* (Araneae: Salticidae). *Chromosoma*, Vol.85, No.1 (April 1982), pp. 23-37, ISSN 0009-5915

Matsumoto, S. (1977). An observation of somatic chromosomes from spider embryo-cells. *Acta Arachnologica*, Vol.27, pp. 167-172, ISSN 0001-5202

McKee, B.D. & Handel, M.A. (1993). Sex chromosomes, recombination and chromatin conformation. *Chromosoma*, Vol. 102, No.2, (January 1993), pp. 71-80, ISSN 0009-5915

Mittal, O.P. (1964). Karyological studies on the Indian spiders II. An analysis of the chromosomes during spermatogenesis in five species of spiders belonging to the family Salticidae. *Research Bulletin (N.S.) of the Panjab University*, Vol.15, No.3-4, (December 1964), pp. 315-326, ISSN 0555-7631

Montgomery, T.H. (1905). The spermatogenesis of *Syrbula* and *Lycosa*, with general considerations upon chromosome reduction and the heterochromosomes. *Proceedings of the Academy of Natural Sciences of Philadelphia*, Vol.57, (February 1905), pp. 162-205, ISSN 0097-3157

Oliveira, E.G.; Cella, D.M. & Brescovit, A.D. (1997). Karyotype of *Loxosceles intermedia* and *Loxosceles laeta* (Arachnida, Araneae, Sicariidae): NeoX$_1$ NeoX$_2$ Y sex determination mechanism and NORs. *Revista Brasileira de Genética*, Vol.20, pp.77, ISSN 0100-8455

Oliveira, R.M.; Jesus, A.C.; Brescovit, A.D. & Cella, D.M. (2007) Chromosomes of *Crossopriza lyoni* (Blackwall 1867), intraindividual numerical chromosome variation in *Physocyclus globosus* (Taczanowski 1874), and the distribution pattern of NORs (Araneomorphae, Haplogynae, Pholcidae). *Journal of Arachnology*, Vol.35, No.2, pp. 293-306, ISSN 0161-8202

Painter, T.S. (1914). Spermatogenesis in spiders. *Zoologische Jahrbuecher Abteilung fuer Anatomie und Ontogenie der Tiere*, Vol.38, pp. 509-576, ISSN 0044-5177

Parida, B.B. & Sharma, N.N. (1986). Karyotype and spermatogenesis in an Indian hunting spider, *Sparassus* sp. (Sparassidae: Arachnida) with multiple sex chromosomes. *Chromosome Information Service*, Vol.40, pp. 28-30, ISSN 0574-9549

Pätau, K. (1948). X-segregation and heterochromasy in the spider *Aranea reaumuri*. *Heredity*, Vol.2, No.1, (June 1948), pp. 77-100, ISSN 0018-067X

Pekár, S. & Král, J. (2001). A comparative study of the biology and karyotypes of two central European Zoodariid spiders (Araneae, Zodariidae). *Journal of Arachnology*, Vol.29, No.3, pp. 345-353, ISSN 0161-8202

Pekár, S.; Král, J.; Malten, A. & Komposch, C. (2005). Comparison of natural history and karyotypes of two closely related ant-eating spiders, *Zodarion hamatum* and *Z.italicum* (Araneae, Zodariidae). *Journal of Natural History*, Vol.39, No.19, pp. 1583-1596, ISSN 0022-2933

Pinter, L.J. & Walters, D.M. (1971). Karyological studies I. A study of the chromosome numbers and sex-determining mechanism of three species of the genus *Phidippus* (Aranea: Salticidae, Dendryphantinae). *Cytologia*, Vol.36, No.1, pp. 183-189, ISSN 0011-4545

Platnick, N.I. (July 2011). The world spider catalog version 12.0, In: *American Museum of Natural History*, 07.08.2011, Available from http://research.amnh.org/entomology/spiders/catalog/index.html

Postiglioni, A. & Brum-Zorrilla, N. (1981). Karyological studies on Uruguayan spiders II. Sex chromosomes in spiders of the genus *Lycosa* (Araneae-Lycosidae). *Genetica*, Vol.56, No.1, (July 1981), pp. 47-53, ISSN 0016-6707

Revell, S.H. (1947). Controlled X-segregation at meiosis in *Tegenaria*. *Heredity*, Vol.1, No.3, (December 1947), pp. 337-347, ISSN 0018-067X

Řezáč, M.; Král, J.; Musilová, J. & Pekár, S. (2006). Unusual karyotype diversity in the European spiders of the genus *Atypus* (Araneae: Atypidae). *Hereditas*, Vol.143, (December 2006), pp. 123-129, ISSN 0018-0661

Rowell, D.M. (1985). Complex sex-linked fusion heterozygosity in the Australian huntsman spider *Delena cancerides* (Araneae: Sparassidae). *Chromosoma*, Vol.93, No.2, (November 1985), pp. 169-176, ISSN 0009-5915

Rowell, D.M. (1988). The chromosomal constitution of *Delena cancerides* Walck. (Araneae: Sparassidae) and its role in the maintenance of social behaviour. *The Australian Entomological Society Miscellaneous Publication*, Vol.5, pp. 107-111

Rowell, D.M. (1990). Fixed fusion heterozigosity in *Delena cancerides* Walck. (Araneae: Sparassidae): an alternative to speciation by monobrachial fusion. *Genetica*, Vol.80, No.2, (February 1990), pp. 139-157, ISSN 0016-6707

Rowell, D.M. (1991). Chromosomal fusion and meiotic behavior in *Delena cancerides* (Araneae: Sparassidae). I. Chromosome pairing and X-chromosome segregation. *Genome*, Vol.34, No.4, (August 1991), pp. 561-566, ISSN 0831-2796

Sharma, G.P.; Jande, S.S. & Tandon, K.K. (1959). Cytological studies on the Indian spiders IV. Chromosome complement and meiosis in *Selenops radiatus* Latreille (Selenopidae) and *Leucauge decorata* (Blackwall) (Tetragnathidae) with special reference to XXX0-type of male sex determining mechanism. *Research Bulletin (N.S.) of the Panjab University*, Vol.10, No.1, (March 1959), pp. 73-80, ISSN 0555-7631

Sharp, H.E. & Rowell, D.M. (2007). Unprecedent chromosomal diversity and behaviour modify linkage patterns and speciation potential: structural heterozigosity in an Australian spider. *Journal of Evolutionary Biology*, Vol.20, No.6, (November 2007), pp. 2427-2439, ISSN 1010-061X

Silva, D. (1988). Estudio cariotípico de *Loxosceles laeta* (Araneae: Loxoscelidae). *Revista Peruana de Entomologia*, Vol.31, (December 1988), pp. 9-12, ISSN 0080-2425

Silva, R.W.; Klisiowicz, D.R.; Cella, D.M.; Mangili, O.C. & Sbalqueiro, I.J. (2002). Differential distribution of constitutive heterochromatin in two species of brown spider: *Loxosceles intermedia* and *L. Laeta* (Araneae, Sicariidae), from the metropolitan region of Curitiba, PR (Brazil). *Acta Biologica Paranaense*, Vol.31, No.1-4, pp. 123-136, ISSN 0301-2123

Stávale, L.M.; Schneider, M.C.; Brescovit, A.D. & Cella, D.M. (2011). Chromosomal characteristics and karyotype of Oxyopidae spiders (Araneae, Entelegynae). *Genetics and Molecular Research*, Vol.10, No.2, (May 2011), pp. 752-763, ISSN 1676-5680

Suzuki, S. (1952). Cytological studies in spiders II. Chromosomal investigation in twenty two species of spiders belonging to the four families, Clubionidae, Sparassidae, Thomisidae and Oxyopidae, which constitute Clubionoidea, with special reference to sex chromosomes. *Journal of Science of the Hiroshima University, Series B, Division 1*, Vol.13, pp. 1-52, ISSN 0368-4113

Suzuki, S. (1954). Cytological studies in spiders III. Studies on the chromosomes of fifty-seven species of spiders belonging to seventeen families, with general considerations on chromosomal evolution. *Journal of Science of the Hiroshima University, Series B, Division 1*, Vol.15, pp. 23-136, ISSN 0368-4113

Wagner, J. (1896). Einige Beobachtungen über die spermatogenese bei den spinnen. *Zoologischer Anzeiger*, Vol.19, pp. 180-190, ISSN 0044-5231

Wallace, L.B. (1900). The acessory chromosome in the spider. *Anatomischer Anzeiger*, Vol.18, pp. 327-329, ISSN 0003-2786

Wallace, L.B. (1905). The spermatogenesis of the spider. *Biological Bulletin*, Vol.8, No.3, (February 1905), pp. 169-184, ISSN 0006-3185

Wallace, L.B. (1909). The spermatogenesis of *Agalena naevia*. *Biological Bulletin*, Vol.17, No.2, (July 1909), pp. 120-161, ISSN 0006-3185

Wang, Z. & Yan, H.M. (2001). Technique of chromosome on spider blood cell. *Chinese Journal of Zoology*, Vol.36, pp. 45-46, ISSN 0250-3263

White, M.J.D. (1940). The origin and evolution of multiple sex-chromosome mechanisms. *Journal of Genetics*, Vol.40, No.1-2, (May 1940), pp. 303-336, ISSN 0022-1333

Wise, D. (1983). An electron microscope study of the karyotypes of two wolf spiders. *Canadian Journal of Genetics and Cytology*, Vol.25, pp. 161-168, ISSN 0008-4093

6

Dynamics of Cellular Components in Meiotic and Premeiotic Divisions in *Drosophila* Males

Yoshihiro H. Inoue, Chie Miyauchi, Tubasa Ogata and Daishi Kitazawa
Insect Biomedical Research Center, Kyoto Institute of Technology,
Japan

1. Introduction

Meiosis in higher organisms is programmed as a part of gametogenesis. It should be discussed separately from a yeast meiosis that initiates in response to extracellular nutrient conditions. In this chapter, we would like to introduce current knowledge including our new findings about dynamics of meiosis in *Drosophila* males. At first, we will introduce general views of cell divisions and cell growth during *Drosophila* spermatogenesis as we illustrated in Fig. 1A. At the tip of the testis in adult *Drosophila* males, several germline stem cells (GSCs) are surrounding next to a cluster of smaller hub cells. The GSCs receive a signal to maintain their multi-potential stem cell characteristics secreted from the adjoining hub cells. A secreted protein encoded by the *unpaired* gene acts as a ligand of the maintenance signal (Kiger, *et al.*, 2001, Fuller & Spradling, 2007). The insulin-like peptides in hemolymph first induce the cell cycle progression of the male GSCs from G2 phase to M phase (Ueishi et al., 2009). Between two daughter cells derived from asymmetric division of the GSC, a proximal daughter cell exclusively receives the Unpaird signal and becomes a self-renewed GSC. The other distal daughter cell leaves the niche and begins to differentiate as a spermatogonium. The spermatogonium then undergoes four cell cycles and generates a 16 cell unit known as a cyst (as a review Fuller, 1993). In ever mitosis, all spermatogonia within a cyst undergo cell divisions synchronously. Cytokinesis in spermatogonia mitoses as well as in following meiotic divisions terminates incompletely. The cleavage furrow ingression is arrested at the middle of cytokinesis and then the contractile rings transform into cytoplasmic bridges called ring canals (Hime et al., 1996). These 16 spermatocytes synchronously enter a growth phase during which the cells remarkably increase in volume by up to 25 times. Following completion of the enormous cell growth, primary spermatocytes carry out two consecutive meiotic divisions. A cyst of 16 spermatocytes gives rise to 64 spermatids simultaneously as a consequence of meiotic divisions. Every spermatid in a cyst at onion stage contains equally sized nucleus and Nebenkern that is a single large aggregate derived from mitochondria. This should be achieved as a consequence of proper chromosome segregation and cytokinesis in germ line cells as well as equal partition of mitochondria (Castrillon et al., 1993, Ichihara et al., 2007).

Next, we describe that the meiotic cells perform dynamic alterations of their cellular components as meiosis progress. If one would like to examine alterations of intracellular structures such as cytoskeletons and organelles during cell division, *Drosophila* male meiotic

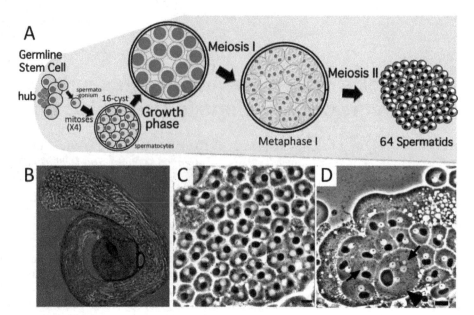

Fig. 1. A. Illustration of premeiotic cell division, cell growth of spermatocytes and meiotic divisions in testis. B. A micrograph of *Drosophila* testis. C, D. A phase contrast micrograph of spermatids at onion stage from wild-type (C) and *orbit[7]* mutant males. Each arrow indicates a smaller nucleus derived from unequal chromosome segregation. Arrowhead shows a larger spermatid containing a large Nebenkern and four nuclei, suggesting a failure of cytokinesis in both meiotic divisions. Bars, 10 µm. Panel C and D were reprinted from Inoue et al., (2004)

cells have several advantages. In *Drosophila*, good cultured cell lines that proliferate well in a standard culture condition are also available (Rogers & Rogers, 2008). However, their cell size, particularly cytoplasmic volume, is much smaller than that of mammalian cells. This is a disadvantage in examination of cellular components during cell division. Spermatocytes, on the other hand, achieve distinct cell growth before initiation of first meiotic division. The primary spermatocyte is the largest diploid cells among proliferative cells in *Drosophila* whole body. Thus, one can easily perform detailed observation of cellular structures in dividing cells using optical microscopes. In *Drosophila melanogaster*, well-advanced and sophisticated genetic techniques are available. Meiotic defects in chromosome segregation and in cytokinesis appear in cellular organization of spermatids just after completion of 2nd meiotic division. By observation of such early spermatids, one can easily find out even subtle meiotic abnormalities. Many of genes essential for cytokinesis have been identified by this method, as we discuss later. Furthermore, if a loss of microtubule integrity or its dynamics would have occurred in normal cultured cells, their cell cycle progression should be arrested before metaphase. Therefore, it is hard to examine how microtubules would influence later processes of cell divisions in the somatic cells. As spermatocytes, on the other hand, are less sensitive to microtubule abnormalities at microtubule assembly checkpoint before metaphase. A colchicine treatment of spermatocytes causes a delay of meiotic cell cycle but it does not make a cell cycle arrest (Rebollo & Gonzalez, 2000). One can, therefore, examine a role of microtubule-related genes in cytokinesis without arresting cell cycle. This is another great advantage of male meiotic cells in cell division studies.

We and other groups have established systems to facilitate dynamics of chromosomes or microtubules by expression of proteins with GFP fluorescence tag (Clarkson & Saint, 1999, Inoue et al., 2004). We describe behavior of chromosomes or chromatids in meiotic divisions and summarize about distribution of homologous chromosomes after premeiotic DNA replication to prophase I. And further, we can also observe other cellular components such as microtubules, actin filaments, endoplasmic reticulum, Golgi apparatuses or mitochondria during male meiosis by a simultaneous expression of proteins with different fluorescence tags. Not only chromosomes but these cellular components also perform dynamic distribution and are equally partitioned in *Drosophila* meiosis. In addition to meiotic divisions, we will also describe here on premeiotic mitoses to generate meiotic spermatocytes. As those cell division, growth and differentiation seen in *Drosophila* spermatogenesis are well conserved among higher eukaryotes, we believe that readers studying on other organisms should also be interested in dynamics of meiotic and premeiotic divisions in *Drosophila* males.

2. Chromosome dynamics in *Drosophila* male meiosis

2.1 Characteristics of chromosome behavior in male meiosis I

Unlike female meiosis, male meiosis in *Drosophila melanogaster* is unique in the following aspects. No chiasmata formation indicating crossing over is observed in male meiosis. Another consistent observation that synaptonemal complex is not formed at prophase I were also reported (Rasmussen, 1973). Thus, meiotic recombination does not occur in *D. melanogaster* males. Even so, the synapsis formation between homologous chromosomes, its maintenance until metaphase I and bivalent disjunction at anaphase I takes a place properly. In past studies, primary spermatocytes dissected from testis were incubated in a culture condition and living

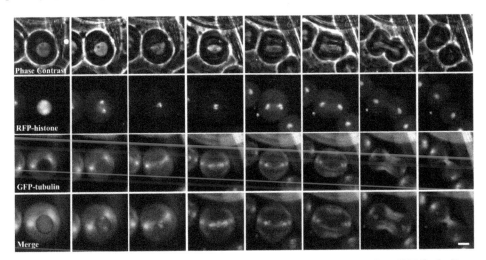

Fig. 2. Time-lapse observation of a living primary spermatocyte expressing GFP-βtubulin and mRFP-Histone2Av simultaneously. Selected fluorescence images of the primary spermatocyte showing chromosomes (red in bottom row, white in second row), microtubules (green in bottom, white in third) and the corresponding phase contrast images to observe cleavage furrow and ER-based cell organelles inside (top). Bar, 10 μm.

spermatocytes were observed under Phase contrast or Differential Interference Contrast (DIC) microscopy equipped video camera (Church & Lin, 1985). In order to trace chromosome behavior together with cell division apparatuses, we have induced simultaneous expression of histone 2Av and βtubulin fused with GFP and mRFP tag, respectively. Time-lapse observation of two kinds of fluorescence makes it possible to examine chromosome behavior such as condensation, congression and segregation as alteration of spindle microtubules (Fig.2).

2.1.1 Chromosome dynamics at meiotic prophase I

Vazquez and colleagues described the distribution of homologous chromosomes at premeiotic stage in primary spermatocytes (Vazquez, et al., 2002). They had established fly stocks in which LacO sequences from E. coli have been inserted into specific sites of Drosophila chromosomes. They then induced expression of GFP-LacI repressors to visualize a distribution of two homologous chromosomes within a nucleus. In spermatocytes at G1 phase just after a completion of 16-cell cyst, homologous chromosomes originally distribute closely each other. This is a characteristic known as a somatic pairing commonly observed in insect cells. This distribution of homologous chromosomes is maintained until meiotic prophase through premeiotic DNA replication. What is a molecular mechanism to fulfill the stable bivalent formation in male meiosis? It has been shown that the homologous chromosomes form a synapsis using specific chromosome regions so-called pairing sites. Church and Lin (1985) proposed that the pairing and separation of sex chromosome X-Y bivalents is controlled by a different mechanism from that of autosome bivalents. It was reported that X-Y pairing depends on a specific pairing site on heterochromatin region of sex chromosomes (Cooper, 1964, McKee and Karpen, 1990), while regions required for autosome pairing distribute across the euchromatin region (Ashburner et al., 2004). The X-Y pairing site seems to correspond to the rDNA gene localized in heterochromatin on X and Y chromosomes. The intergenic spacer regions of rDNA gene clusters play a role in X-Y pairing independent on nucleolus formation (McKee et al., 1992). On the other hand, there seems to be no pairing sites in heterochromatin regions of autosomes (Yamamoto, 1979, Ashburner et al., 2004). The autosome bivalents can be paired along entire chromosome regions. Trans-acting factors required for proper execution of meiotic chromosome pairing has also been identified. It was recently reported that a loss of a chromosomal passenger protein, Australin results in defects in chromosome segregation in Drosophila male meiosis (Gao et al., 2008). As the Australin is exclusively required for male meiosis, these genetic data is consistent with evidences that chromosome segregation in male meiotic divisions is regulated by different mechanisms from that in females. And further, mutations of tef gene encoding a chromatin binding protein with Zn finger disrupted autosomal pairing without affecting X-Y pairing and segregation in male meiosis I (Tomkiel et al., 2001). Common factors essential for the pairing of all chromosomes have been also identified. The mod(mdg4) encoding a chromosome binding protein required for segregation of all chromosomes in male meiosis (Soltani-Bejnood et al., 2007). These genetic data suggest that some of molecular mechanisms in meiotic chromosome segregation are different from each other in females and males but others are common in both sexes.

2.1.2 Chromosome movement in prometaphase I to telophase I

Church and Lin (1985) examined chromosome movement in meiosis I by time-lapse video microscopy. They observed unpredictable movement and frequent reorientation at prometaphase-like stage before the bivalents achieve a stable bipolar orientation. They described that *Drosophila* primary spermatocyte does not form a well-defined metaphase plate. We have observed four foci of GFP-Histone 2Av corresponding to bivalents between two major autosomes, X-Y chromosomes and tiny 4th chromosomes in living primary spermatocytes at prometaphase I (Fig. 2). *Drosophila melanogaster* cells contain 3 major chromosomes and one small chromosome 4. Bivalents of smaller chromosomes are not usually observed because of overlapping with major chromosomes. Those three foci appear to congress into a single chromosome mass at the middle of the primary spermatocytes. Savoian and colleagues found that each chromosome carried out one or more rapid poleward movement with an average velocity of 11.2 ± 1.2 µm/min until bipolar kinetochore attachment (Savoian et. al., 2000, Savoian et al., 2004). However, it is not necessary for anaphase initiation to make all bivalents align at the cell equator. It was also revealed that state of microtubule assembly is surveyed at the M-phase checkpoint, although the checkpoint at male meiosis is less strict than that in somatic cells (Rebollo & Gonzalez, 2000). Past studies reported that chromosome movement at Anaphase I was highly irregular in velocity (Church and Lin, 1985). Savoian and colleagues described that Anaphase I takes 8 ± 1 min and that the chromosomes moved polewards at 1.9 ± 0.1 µm/min after dyad disjoining.

3. Dynamics of cytoskeletons and cell organelles as a progression of male meiosis

3.1 Dynamics of cytoskeletons in spermatocytes

It is well known that cytoskeletons perform dramatic changes in their structures during cell divisions. In *Drosophila* germline cysts, the most characteristic future concerning cytoskeletons in spermatocytes is a presence of fusome that is a germline specific cell organelle (Hime et al., 1996). This is the F-actin based branching structure containing alpha-Spectrin, ß-Spectrin, adducin-like protein (Hu-li tai shao) and Ankyrin. The fusome traverses cytoplasmic bridges to connect clonally related spermatogonia and premeiotic spermatocytes within a cyst. Male germline stem cells and early spermatogonia contain a spherical type of the fusome, called a spectrosome (Yamashita et al., 2003, Wilson, 2005). During the four rounds mitoses of spermatogonia, the fusome forms one large branched structure that extends though the ring canals into all the cells within a cyst. Before meiotic division I, the characteristic branched structure of fusome has disappeared and its fragmented remnants appear during meiotic divisions. And then similar branched structures devoid of F-actin appear to penetrate post-meiotic ring canals (Hime et al., 1996). It is possible to speculate that these fusome structures play a role in connecting individual spermatogonia with each other and in determination of their division axis as we discuss later.

3.1.1 Actin cytoskeletons

At late anaphase, a contractile ring consisting of F-actin and myosin II is constructed on the middle of cell cortex in spermatocytes. Myosin II moves along the F-actin filaments by using the free energy of ATP hydrolysis. Shrinking of the ring constricts the cell membrane to form a cleavage furrow (Egger et al., 2006). Cytokinesis in spermatogonia mitoses as well as in meiotic divisions terminates incompletely. The cleavage furrow ingression is arrested at the middle of cytokinesis and then the contractile rings transform into cytoplasmic bridges called ring canals.

F-actin or Myosin II is no longer found in the ring canals of spermatocytes but instead large amount of phospho-tyrosin, anillin and septins are contained in the bridge architecture (Hime et al., 19996). It has been revealed that the position of contractile ring formation is determined by central spindles (as a review, Goldberg et al., 1998). The bundle of interdigitated microtubules emanating from both spindle poles formed between the separating homologous chromosomes (Inoue et al., 2004). Interestingly, the actin-based contractile ring shows a cooperative interaction with the central spindle microtubules. Both mutations of *chic* gene encoding a profillin homolog essential for actin polymerization and treatment of testis cells with cytochalasin B showed a similar cytokinesis phenotype. In both cases, not only a formation of the actin-based contractile ring but the central spindle microtubules also failed to be constructed (Giansanti et al., 1998). On the contrary, for example, mutations for *orbit* gene encoding a microtubule-associated protein caused a sever disruption of central spindle microtubules and they also resulted in a failure of cytokinesis devoid of the F-actin rings (Inoue et al., 2004). These results strongly suggest that the central spindle microtubules and the contractile ring consisting actomyosin are interdependent structures, at least during cytokinesis in male meiosis. A molecular mechanism to link these intracellular structures remains to be uncovered. It should be necessary to identify a key molecule(s) to interact directly with both the central spindle microtubules and components of the contractile ring.

3.1.2 Microtubule structures

Drosophila primary spermatocytes possess characteristic microtubule structures for a preparation of meiotic divisions. Particularly, well-developed astral microtubules become to be prominent as initiation of meiosis I in primary spermatocytes. The astral microtubules are easy to recognize from spindle microtubules that are formed in the nuclear space surrounded by multiple membrane layers known as parafusorial membranes (Tates, 1971, Fuller, 2004, Bonaccorsi et al., 2000). To visualize astral and spindle microtubules and to examine the dynamics of microtubule behaviour in living primary spermatocytes, Inoue and colleagues used a transgenic line ubiquitously expressing GFP-tagged β-tubulin (Inoue et al., 2004). They performed simultaneous observation of multiple cellular structures by DIC microscopy as well as fluorescence microscopy. We presented here several selected figures of a primary spermatocyte undergoing meiosis I simultaneously expressing mRFP-histone 2Av and GFP-βtubulin (Fig. 2). As chromatin inside of nucleus has unevenly distributed at late prophase (t=0 minute), microtubules around both spindle poles become to be prominent. At prometaphase I, four condensed bivalent chromosomes would be observed within a nucleus in which nuclear membrane seems to be intact (t=5, under phase contrast microscope). Only two dyads seem to be in focus in Fig. 2. Developing asters have moved around nuclear membrane as to reach at opposite poles. Then, at 20 minutes later, all chromosome complements congress at the centre of the bipolar spindle structure. At this stage, nuclear morphology has already disintegrated and spindle microtubules free to elongate into inside of nuclear space. The kinetochore microtubules seem to capture the chromosomes and put them in the centre. At onset of anaphase I (t=50), A multilayer of nuclear membrane see as phase-dark structure surrounding around nuclear space separate spindle microtubules including thick kinetochore microtubules from well developed astral microtubules nucleated around each spindle pole. Two populations of central spindle microtubules appear after disjunction of bivalents (t=60 to 70, Fig. 1). A peripheral set of the microtubules elongating from spindle pole regions become more dynamic as if they look for the cytoplasm towards the cell equator (Inoue et al., 2004). Another set of the microtubule bundles corresponding spindle microtubules is localised interiorly at the middle of the cell. The peripheral microtubules from opposite poles met together at equator and form bubble-like structures protruding outwards (t=60). The interior and most of the peripheral central spindles then are released from each pole and they formed independent bundles at the equator (Inoue et al., 2004, Savoian et al., 2004). Furrow ingression was observed soon after the peripheral microtubules from both poles contacted the cell cortex. Thus, we speculate that interaction between the peripheral microtubules and the cell cortex plays a role in determination where and when cleavage furrow ingression initiate in male meiosis.

3.2 Mitochondria

It had been believed that inheritance of cytoplasmic organelles such as mitochondria is achieved passively as a consequence of equal cytokinesis (Shima and Warren, 1998). However, it has been described that mitochondria are transmitted toward a daughter cell by the active transport system in fission yeast mitosis (Yaffe et al., 2003). In *Drosophila* male meiosis, the mitochondria line up along nuclear membranes and are equally divided to two daughter cells at each division of *Drosophila* male meiosis (Fuller, 1993). Ichihara and colleagues examined carefully distribution of mitochondria during meiotic divisions in *Drosophila* males (Fig. 3 from Ichihara et al., 2007). At a beginning of prophase (Fig. 3A),

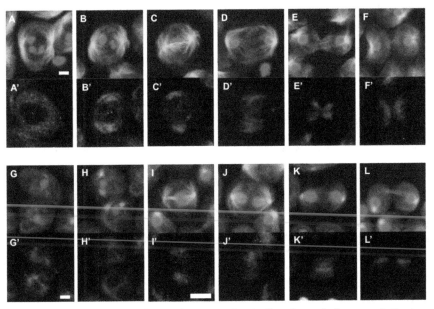

Fig. 3. Distribution of mitochondria with aster and spindle microtubules at meiotic stages in male meiosis I and II. Normal spermatocytes undergoing meiosis I (A-F) or II (G-L) were stained to visualize microtubules (green), mitochondria (red) and chromosomes (blue). Prophase (A, H), prometaphase (B), metaphase (C, I), anaphase A (D, J), anaphase B (E, K), telophase (F, L) and interphase (G), respectively. Bars, 10 μm. These all panels were reprinted from Ichihara et al., (2007).

mitochondria are homogenously distributed throughout cytoplasm. As aster microtubules are developing at prometaphase (Fig. 3B), mitochondria are expelled from inside of the asters and then assembled toward plus ends of aster microtubules. At metaphase, the mitochondria are clustered at the equator of a peripheral cytoplasmic region between facing two asters (Fig. 3C). As central spindle microtubules are formed between separating sister chromatids at anaphase (Fig. 3D), these mitochondria distribute along central spindles as if to decorate the microtubule structures. The mitochondria dispersed into cytoplasm (Fig. 3F), as the central spindles disintegrated. At the end of meiotic division II, mitochondria release from midbody microtubules and assemble to form single large aggregate called Nebenkern (Fig. 1). This transition of mitochondria is regulated by an ordered mode rather than a stochastic partitioning strategy. As spermatozoa require a consumption of higher amount of ATP for the motility, existence of regulation to facilitate equal inheritance of mitochondria in male meiosis may be of advantage that can produce a large number of spermatozoa with homogeneous quality at the same time.

3.3 Golgi stacks

During cell division, it was shown that some cell organelles lose their function and are fragmented by modification of their key components (Rabouille and Jokitalo, 2003; Rabouille and Klumperman, 2005). Golgi apparatus in mammalian cells is consisting of flattened membrane-bound compartments. The stacks interconnect each other and form a single large organelle so-called Golgi ribbon beside nucleus. The mammalian Golgi apparatus undergoes disassembly at onset of mitosis and reassembles into the ribbon structure at telophase (Misteli, 1997). A major difference in the Golgi organization between the mammalian and *Drosophila* cells is that a single Golgi ribbon as seen in mammals is not observed in *Drosophila*. Instead, multiple sets of smaller tER-Golgi unit that is a complex consisting of tER site and a piece of Golgi stacks dispersed throughout the cytoplasm (Kondylis & Rabouille, 2009). As membranous intracellular structures such as parafusorial membranes are well developed in spermatocytes (Tates 1971, Fuller, 2003), it is interesting to take up a dynamic behavior of membrane-based organelles in premeiotic and meiotic cells. It is considered that the Golgi stacks in *Drosophila* cultured cells also display a cycle of disassembly and reassembly during mitosis (Kondylis et al., 2007). We also observed carefully distribution of Golgi stack components visualized by immunostaining with antibody against a cis-Golgi protein, GM130 at several meiotic stages (Fig. 4). After spermatocytes initiate meiosis I, the Golgi-derived vesicles seem to increase in the number (compare Fig. 4A with B). The Golgi-derived vesicles were first uniformly distributed throughout the cytoplasm (Fig. 4A). They were then assembled into two groups containing similar amounts of Golgi vesicles prior to chromosome segregation (Fig. 4B). The Golgi vesicles were accumulated around each spindle pole until mid-telophase. At the end of cytokinesis after stage E in Fig. 4, the Golgi vesicles become to be redistributed throughout cytoplasm. These observations remind us equal partition of mitochondria in male meiosis described above. It might be possible to interpret that the Golgi-derived vesicles would be also partitioned equally between two daughter cells in male meiotic divisions. However, it was reported that vesicles containing another Golgi component, Rab11 was concentrated on the Golgi stacks and at the nuclear envelope from prophase to metaphase and that the Golgi-derived vesicles were subsequently accumulated at the cell poles (Giansanti, et al., 2007). At mid-telophase, the vesicles became to concentrate at cell equator possibly to contribute new membrane insertion in cytokinesis. Accumulation of Rab11-containing Golgi vesicles at each cell pole may be no more than storage of vesicles required at cleavage furrow site. Further examinations using other Golgi-related vesicles should be necessary to conclude significance of Golgi-derived vesicle distribution during male meiotic divisions.

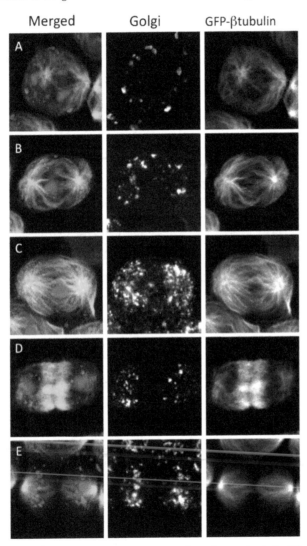

Fig. 4. Immunostainig of primary spermatocytes undergoing meiosis I with antibody against the cis-Golgi protein. Interphase or prophase I (A), prometaphase I (B), metaphase I (C), late anaphase I (D), telophase I (E). (Red) immunostaining with anti-GM130 antibody that recognizes a cis-Golgi protein. (Green) microtubules visualized by GFP-βtubulin, (blue) DNA staining. Prophase (A, H), prometaphase (B), metaphase (C, I), anaphase A (D, J), anaphase B (E, K), telophase (F, L) and interphase (G), respectively. Bars, 10μm

3.4 Endoplasmic Reticulum- based structures

Expression of Rtnl1 proteins residing predominantly in endoplasmic reticulum (ER) fused with GFP tag allowed us to observe its dynamics in primary spermatocytes. The ER-residing proteins are distributed on phase-dense structures observed around the nuclear space under phase contrast microscope (Miyauchi and Inoue, unpublished). This observation showed that the ER constructs distinct intracellular structures in spermatocytes during male meiotic divisions. Dorogova and colleagues have reported that ER network has distinctive reticular morphology immediately before initiation of male meiosis (Dorogova et al., 2009). The ER forms concentric circles on the outside of nuclear membrane at meiotic prophase. This ER network then dramatically changes as follows. As a progression of meiotic divisions, the ER networks form multi-layers of branchless membranous sheets appeared as phase-dense structure around nuclear space (Fig. 2 upper row). After prometaphase to telophase, the membranous structure develop very well around astral microtubules. This structure possibly corresponds to cellular organelles known as astral membrane in *Drosophila* embryos. This membranous organelles in embryo seem to be mainly consisted of ER sheets (Bobinnec et al., 2003). The reticular ER structure elongates along a direction of division axis and it transforms into multiple layers of membranes known as spindle envelopes, which surround the nuclear space. These two ER-based structures, astral membranes and spindle envelopes are especially characteristic in male meiotic cells. And they seem to be closely associated with astral microtubules and central spindle microtubules, respectively. These meiotic configurations restore to the reticular distribution around nucleus in next interphase. It is reasonable to speculate that these ER-based structures would interact with other cellular components and may have a role to facilitate dynamic changes of cellular components.

4. Identification of genes essential for proper execution of chromosome segregation and cytokinesis in male meiosis

Drosophila male meiosis provides us an advantage in examination of chromosome segregation and cytokinesis (Fuller, 1993; Maines and Wasserman, 1998). After a completion of meiotic divisions, every spermatid in a cyst consisting 64 cells contains equally sized single nucleus and a Nebenkern. This should be achieved as a consequence of proper chromosome segregation and cytokinesis in germ line cells. This is a quite convenient and sensitive method to find out those cell division defects in either or both meiotic divisions in living spermatids (Castrillon et al., 1993). Mutants for either chromosome segregation or cytokinesis or even both in male meiotic divisions could easily be distinguished by the spermatid morphology at onion stage. The evidences that the volume of spermatid nuclei is proportional to numbers of chromosome complements allow us to find out defects in chromosome segregation on the base of a presence of different sized spermatids (Gonzalez et al., 1989). A failure of cytokinesis results in generation of spermatids carrying abnormally large Nebenkern and multiple nuclei. Another advantage of *Drosophila* male meiosis is its less strict checkpoint to monitor spindle assembly before metaphase. Even if meiotic cells carry spindle defects, they only delays anaphase onset rather than arresting the cell cycle before metaphase as observed in somatic cells. Therefore, one can examine influence of microtubules on later stages after microtubule assemble checkpoint such as chromosome segregation at anaphase or cytokinesis. After picking up mutant candidates by spermatid morphology, one can perform further careful examination in primary spermatocyte that is the largest proliferative cell. Using this convenient examination system, many of genes required for meiotic divisions have so far been identified. However, disadvantage of the cell type is that mutants for essential genes cannot usually be examined due to their earlier lethality. Hypomorphic mutations that can overcome the lethality at earlier developmental stages are useful for this purpose. And furthermore, we and other group have recently succeeded to perform knockdown experiments by induction of dsRNA for genes in premeiotic spermatocytes using the Gal4/UAS system (Goldbach et al., 2010 , Kitazawa et al., submitted). We put here a partial list of the genes required for meiotic cytokinesis in *Drosophila* male with especial focus on genes involved in a formation of central spindle microtubules and contractile ring (Table 1). The responsible genes for cytokinesis mutants includes genes encoding factors involved in dynamics of actin filaments, microtubules, motor proteins along cytoskeletons and so on. It is well known that spindle microtubules and contractile rings consisting actomyosin display mutual dependency in construction and function (Giansanti et al., 1998). To understand molecular mechanism of crosstalk between these two cytoskeletons, these *Drosophila* mutants and dsRNA stocks should be valuable genetic tools. These cytokinesis phenotypes appeared in *Drosophila* male meiosis, however, have not been completely confirmed by knockdown experiments in S2 cultured cells (Somma et al., 2002). For example, knockdown of *chic*, *fwd* and *klp3A* that have previously identified as essential genes for male meiosis did not impair cytokinesis of S2 cells treated with dsRNA. It is possible to speculate that a requirement of these gene products might differ from each cell type. It is important to continue comparative studies between somatic cells and meiotic cells for understanding general mechanisms to control cell divisions.

Protein name	Symbol	Biochemical activity	Cell function	orthologue
abnormal spindle	asp	microtubule-associated protein (MAP)	microtubulesorganization & cytokinesis	ASPM(mammal)
Adenomatous polyposis coli homolog 2	Apc2	adenomatous polyposis coli homolog	centrosome localization in Germline Stem cells (GSCs)	Apc
Apc-like	Apc	adenomatous polyposis coli homolog	centrosome localization in GSCs	Apc
Anillin	scra	component of the contractile ring	cytokinesis	Anillin
arrest	aret	RNA-binding protein	regulation of mRNAs involved in gametogenesis	
asterless	asl	centriolar or PCM protein	centrosome organization	
aurora B	ial	Serine/Threonine kinase	chromosome condensation and cytokinesis	Aurora B
auxilin	aux	Serine/Threonine kinase	cytokinesis	cyclin G associated kinase

Table 1. Proteins involved in organization of microtubules and actin filaments in *Drosophila* male meiosis.

Protein name	Symbol	Biochemical activity	Cell function	orthologue
bag-of-marbles	bam	Serine/Threonine kinase	spermatogonia deivision	
james bond	bond	Serine/Threonine kinase	contractile ring assembly	GNS1/SUR4 (yeast)
Btk family kinase at 29A	Btk29A	protein-tyrosine kinase	maintaining the equilibrium between G- and F-actin	Btk
Bub1	bub1	Serine/Threonine kinase	microtubule assembly checkpoint	Bub1
Centrosomin	cnn	pericentrosomal matrix protein	centrosome organization	CDK5 RAP2(mammal)
Chickadee	chic	profilin homolog	regulates polymerization of the actin cytoskeletone	Profilin
CIN85 and CD2AP orthologue	cindr	adaptor protein	links cell surface junctions and adhesion proteins	CIN85, CD2AP
Cortactin	cortactin	cytoskeletal component	stimulate actin polymerizationx	cortactin
courtless	crl	ubiquitin conjugating enzyme	male meiosis	Ubiquitin-conjugating enzyme, E2
diaphanous	dia	Formin homology domain protein	cytokinesis	mDia(mammal)
Dynamin related protein 1	drp1	Dynamin family	mitochondrial fusion	Dynamin
effete	eff	UbcD1	germline stem cell maintenance	
exo84	exo84	Cullin repeat	contractile ring constriction	exo8
Fmr1	Fmr1	RNA binding		FMR1
four wheel drive	fwd	Phosholipid kinase	contractile ring constriction	1-phosphatidylinositol 4-kinase
four way stop	fws		contractile ring constriction	Cog5
Fps oncogene analog	Fps85D	protein tyrosin kinase	protein tyrosin kinase cytoskeletal reorganization	Fps
sec8	sec8	exocyst complex component	contractile ring constriction	sec8
gilgamesh	gish	Ser/Thr kinase sperm indivisualization	Ser/Thr kinase sperm indivisualization	casein kinase
Kinesin-like protein at 61F	klp61F	kinesin-5 family protein	microtubule crosslinking and sliding activities	Eg5
Kinesin-like protein at 3A	klp3A	kinesin-4 family	cytokinesis	KIF4A
loquacious	loqs	RNA binding miRNA processing	germline stem cell maintenance	TARBP2
mei-s332	mei-s332	Shugoshin related	sister chromatid cohesion	Shugoshin related

Table 1. Proteins involved in organization of microtubules and actin filament in *Drosophila* male meiosis. (Continuation)

Protein name	Symbol	Biochemical activity	Cell function	orthologue
Myt1	Myt1	protei kinase	meiosis and spermatid differentiation	Myt1
off-schedule	ofs	eIF4G2 translation initiation	meiosis	elF
orientation disrupter	ord	unknown	meiotic sister chromatid cohesion	
parkin	park	E3 Ubiquichin ligase	mitochondria organization	parkin
pavarotti	pav	kinesin-6 family	spindle organization and cytokinesis	MKLP1 related
peanut	pnut	septin family	contractile ring formatioin	septin
pebble	pbl	RacGAP50C	cytokinesis	ECT2
pelota	pelo	eRF1 domain protein	male meiosis progression	pelota homolog
pericentrin-like protein	cp309	centriolar proetin	spindle pole organization	pericentrin
plk4	plk4	Ser/Thr kinase	centriole duplication	plk4
polo	polo	Ser/Thr kinase	cytokinesis central spindle & contractile ring	plk
Rab-protein 11	Rab11	small GTPase	cytokinesis membrane traffic	Rab11
Rac1	Rac1	small G protein	actin filament organization	Rac1
rhomboid-7	rho-7	rhomboid family proteinase	mitochondrial fusion in spermatogenesis	PARL protease
shutdown	shu	FK506-binding protein domain	germline stem cell regulation	
Spageghetti squash	sqh	Myosin regulatory light chain	formation of central spindle & contractile ring	Myosin regulatory light chain
Spectrin a	aSpec	cytoskeletal protein	fusome organization	Spectrin a
spindle assembly abnormal 6	sas-6	centriolar proetin	centrosome duplication	Sas6
sticky	sti	Rho effector kinase	cytokinesis	citron
subito	sub	kinesin-like protein	meiosis spindle organization	klp9
Syntaxin 5	Syx5	t-SNARE homology	golgi traffic assembly	Syntaxin 5
Transforming growth factor beta at 60A	gbb	TGF ligand	regulate germ line stem cells and other process	TGFb
twinstar	tsr	cofilin	meiosis and spermatid differentiation other process	cofillin
vibrator	vib	Phosphatidylinositol transfer protein	contractile ring constriction	
zipper	zip	myosin heavy chain II	cytokinesis formation of central spindle & contractile ring	Myosin II

Table 1. Proteins involved in organization of microtubules and actin filament in *Drosophila* male meiosis. (Continuation)

Information about each *Drosophila* gene can be obtained from the flybase by linking to the following URL: http://flybase.org/.

5. Signaling pathways to control remarkable cell growth of spermatocytes before meiotic initiation

5.1 Cell growth of premeiotic spermatocytes

The *Drosophila* spermatocytes increase in size up to 25 times during 90 hours after premeiotic DNA replication. The cell growth of premeiotic spermatocytes is the largest cell growth among that seen in *Drosophila* proliferative cells. The growth phase of primary spermatocytes is clarified as S1 to S6 stages (Cenci et al., 1994, Bonaccorsi et al., 2000). After four rounds of spermatogonia mitosis to form a 16-cell cyst, premeiotic DNA replication takes a place at first S1 stage. S2 to S6 stage corresponds to extended G2 phase of cell cycle. The later S6 stage is possibly overlapping to meiotic prophase. In later S2 phase known as polar spermatocyte, mitochondria increase in number and form a cluster at opposite side of the nucleus, while nucleus is dislocated from central position. As the cell volume of spermatocytes has gradually increased after the S2 stage, a plenty of genes including testis-specific gene, βtub85D are highly transcribed at the growth stages. Male fertility factors such as KS-1 are highly expressed so that its transcripts become to be visible under phase contrast microscope (Bonaccorsi et al., 2000). Because chromosomes are highly condensed during meiosis and after later spermatogenesis, most of gene products required for meiotic divisions and later spermatogenesis should be expressed in primary spermatocytes before meiotic initiation. The extreme cell growth may be achieved in coordination with enhanced expression and accumulation of proteins required for later cell divisions and development.

5.2 Growth factors and signaling cascades to induce spermatocyte growth

As described above, the *Drosophila* spermatocytes have achieved most distinctive cell growth up to 25 times after premeiotic DNA replication. What is a molecular mechanism to induce such an enormous spermatocyte growth? Insulin-like peptides (ILPs) play an important role in induction of somatic cell growth (Brogiolo, et al., 2001, Ikeya et al., 2002). Ueishi and colleagues reported that a loss of ILPs by specific apoptosis induction to insulin-producing cells results in reduced growth of spermatocytes, suggesting that the spermatocyte cell growth is required for ILPs (Rulifson et al., 2000, Ueishi et al., 2009). They further showed that an accumulation of active Akt form phosphorylated by its upstream factor, PDK1 in the growing spermatocytes. A diameter of spermatocytes from mutant males for Insulin Receptor (InR) or IRS orthologue encoded by *chico* gene decreased to 70 % of normal size. These genetic data suggest that the insulin signaling plays an essential role in the remarkable cell growth of spermatocytes (Ueishi et al., 2009). We further examined whether PI3 kinase acting upstream Akt is also involved in cell growth induction. The expression of constitutive active form of PI3 kinase catalytic subunit was induced in spermatogonia to premeiotic spermatocyte stage (Ogata and Inoue unpublished). Such an induced expression results in 14% increase of spermatocytes in diameter. These genetic data strongly suggest that the ILPs and its signaling cascade through PI3 kinase to Akt plays a role in induction of spermatocyte cell growth in *Drosophila*. As mammalian insulin can also activate the Ras-MAP kinase cascade after the Insulin receptor (as a review, Avruch, 1998), we further examine whether Ras signaling cascade acting downstream of Drosophila Insulin Receptor homologue (InR) is also involved in the cell growth of spermatocytes before male meiosis. Constitutively activated mutation for Ras85D, Ras85D^{v12} (Kim et al., 2006) also induced approximately 10 % increase of cell diameter in length (Ogata and Inoue, unpublished). Therefore, these genetic data suggest that both PI3K-

Akt cascade and Ras-MAP kinase cascades acting downstream of InR are essential for induction of the premeiotic spermatocyte growth.

6. Asymmetric division of germline stem cells and directional divisions of spermatogonia before meiosis

At the tip of the testis in adult *Drosophila* males, several germline stem cells (GSCs) can be observed. The GSCs receive a signal to maintain their stem cell characteristics secreted from the adjoining hub cells. A ligand protein encoded by the *unpaired* gene is used as the maintenance signal and the signal is transmitted through the JAK-STAT signaling cascade (Kiger, *et al.*, 2001, Tulina & Matunis, 2001). A proximal cell of the two daughter cells derived from an asymmetric division of the GSC exclusively receives the signal and becomes a self-renewed GSC. For self-renewal and differentiation of GSC daughters, it is crucial to set up cell division axis perpendicular to a cluster of the hub cells (Yamashita et al., 2003). What is a molecular mechanism to set up the spindle axis perpendicular? It was shown that mother centrosome in GSC remains to be positioned at the cell cortex contiguous to the hub cells and that daughter centrosome derived from duplication of the mother in GSCs is released from the cortex and migrates toward an opposite pole (Yamashita et al., 2007). The distal daughter cell derived from a GSC division leaves the niche and differentiates as a spermatogonium.

The spermatogonium then initiates four times of cell cycles to generate a cyst consisting of 16 cells. Every spermatogonium in a cyst undergoes these four rounds of cell division synchronously. The orientation of these spermatogonia divisions rotates at 90 degree in every mitosis. Like GSCs, it is possible to speculate that a daughter centrosome derived from the mother anchored to the fusome is free from a connection with the fusome extended over spamatogonia within a cyst and thus it could migrate toward an opposite pole until prometaphase. In this way, spermatogonia could alternate division orientation at 90 degree in every cell division. Such unusual mitoses may be advantageous to store a cyst containing constant numbers of spermatogonia within a limited space of testis.

7. Future researches

The primary spermatocyte is considered as one of the cells most thoroughly examined about cell division together with a S2 cultured cell. In addition to genetic analyses using hypomorphic mutants viable up to developmental stages in which male meiosis can be observed, targeted knockdown of all most of the *Drosophila* genes currently became to be possible in spermatocytes. We can expect that saturation genetic studies to examine variety of phenotypes should be frequently carried out near future. It should be fruitful for us to collect whole information about cell phenotypes appeared in primary spermatocytes from such large scale knock down experiments. These efforts would certainly increase value of male meiotic cell as a model cell for researches on cell proliferation and growth. As the primary spermatocytes have some specific futures in terms of intracellular structures or cell cycle regulation, it is necessary to perform comparative studies using common cultured cells for confirmation of genetic results obtained in male meiotic cells. For more detailed real-time observation to examine dynamics of multiple cellular components simultaneously during male meiosis, it is important to develop cultured system to make it possible to do longer observation from spermatogonial stage to onion stage of spermatid. We currently succeeded to carry out continuous observation from onset of meiosis to onion stage spermatids. It is also necessary to establish *Drosophila* stocks to induce simultaneous expression of several cellular components fused with different fluorescence tag, GFP, mRFP or CFP. They allow us to perform multi-color time lapse observation that would make it possible to trace alterations of chromosomes, microtubules or actin filaments and other cellular organelles in a single cell. Such a new observation system should stimulate understanding of dynamic feature of male meiosis in *Drosophila*. As a series of cellular events such as cell division, growth, elongation and differentiation in spermatogenesis is considerably conserved between *Drosophila* and mammals (Zhou and Griswold, 2008), we expect that *Drosophila* data would also bring us valuable information to help better understanding of mammalian spermatogenesis.

8. Conclusion

In this chapter, we conclude that *Drosophila* primary spermatocytes undergoing meiosis I is an excellent model cell to examine dynamics of chromosomes and other cellular components such as cell organelles and cytoskeletons. Although chiasmata formation and homologous recombination does not occur in *Drosophila* male meiosis, we can study on more simple chromosome pairing and segregation of homologous chromosomes. We showed a simultaneous observation of chromosomes, microtubules and cell organelles in living primary spermatocytes expressing proteins fused with different fluorescence tags. Microtubules and actin filaments display dynamic alterations as a progression of male meiotic divisions. Furthermore, our observation indicates cell organelles such as mitochondria or golgi foci are also transmitted equally toward two daughter cells dependent on microtubule structures. By examination of hypomorphic mutants or knockdown spermatocytes, it have been shown that a plenty of cell cycle related genes including many novel genes play a important role in male meiotic divisions. Before initiation of meiotic division I, the cells achieve largest extent of cell growth in *Drosophila*. We also discussed molecular mechanisms to induce the distinct cell growth of premeiotic spermatocytes by insulin-like peptides, their signaling pathways and other related pathways showing crosstalk with the insulin cascade. In addition to male meiotic divisions, we briefly referred four round premeiotic divisions of spermatogonia to generate a 16-cell cyst and discussed about its regulatory mechanism. These mitoses are synchronous cell divisions in which spindle axis rotate by 90 degree in every division. Animal meiosis is a part of development programs in gametogenesis. It is basically different from yeast meiosis that can be induced by environment cues. *Drosophila* male provides us a good model to understand common molecular mechanisms to control animal meiosis.

9. Acknowledgment

We would like to acknowledge L. Cooley, E. Mathe, R. Saint, and Bloomington stock center, Drosophila Genetic Resource Center, Vienna Drosophila RNAi Center, for providing fly stocks and This work was partially supported by Grants-in-Aid for Scientific Research on Priority Area and Grants-in-Aid for Scientific Research (C) to Y.H.I.

10. References

Ashburner, M., Golic, K. G. & Hawley R. S. (2004) Male meiosis, *Drosophila. A Laboratory Handbook, 2nd ed.* pp.827-860. Cold Spring Harbor Laboratory Press, ISBN0-87969-706-7, NewYork, USA

Avruch, J. (1998). *Insulin signal transduction through protein kinase cascades Mol.Cell. Biochem.* 182: pp.31–48, 1998. doi:10.1023/A:1006823109415

Bobinnec, Y., Marcaillou, C., Morin, X. & Debec. A. (2003). Dynamics of the endoplasmic reticulum during early development of *Drosophila* melanogaster. *Cell Motil Cytoskeleton.* 54: 217-225. doi: 10.1002/cm.10094

Bonaccorsi, S., Giansanti, M.G., Cenci, G., and Gatti, M. (2000). Cytological analysis of spermatocytes growth and male meiosis in *Drosophila melanogaster.* In: *Drosophila* Protocol. (W. Sullivan, M. Ashburner, R. S. Hawley, eds.) Cold Spring Harbor Laboratory Press, ISBN0-87969-584-4, Cold Spring Harbor, New York, pp 87-109.

Castrillon, D.H., Gonczy, P., Alexander, S., Rawson, R., Eberhart, C.G., Viswanathan, S., DiNardo, S., & Wasserman, S.A. (1993). Toward a molecular genetic analysis of spermatogenesis in Drosophila melanogaster: characterization of male-sterile mutants generated by single P element mutagenesis. *Genetics* 135: 489-505.

Cenci, G., Bonaccorsi, S., Pisano, C., Verni, F., and Gatti, M. (1994). Chromatin and microtubule organization during premeiotic, meiotic and early postmeiotic stages of *Drosophila melanogaster* spermatogenesis. *J. Cell Sci.* 107: 3521-3534.

Cooper, K. W. (1964). Meiotic conjunctive elements not involving chiasmata. *Proc. Natl. Acad. Sci.* 52:1248-1255. doi:10.1073/pnas.52.5.1248

Church, K. & Lin, H. P. (1985) Kinetochore microtubules and chromosome movement during prometaphase in *Drosophila melanogaster* spermatocytes studied in life and with the electron microscope. *Chromosoma* 92: 273-282. doi:10.1007/BF00329810

Drogova, N. V., Nerusheva, O. O. & Omelyanchuk, L. V. (2009). Structural organization and dynamics of the endoplasmic Reticulum during spermatogenesis of Drosophila melanogaster: Studies using PDI-GFP chimera protein. *Biochemistry (Moscow)* :55-61.

Duffy, J. B. GAL4 system in *Drosophila.* (2002) A fly geneticist's swiss army knife. *Genesis* 34:1–15. doi:10.1002/gene.10150.

Eggert, U. S., Kiger, A. A., Richter, C., Perlman, Z. E., Perrimon, N., Mitchison, T. J., & Field, C. M. (2004) Parallel Chemical Genetic and Genome-Wide RNAi Screens Identify Cytokinesis Inhibitors and Targets. *PLoS Biol.* 2: e379. doi:10.1371/journal.pbio.0020379

Fuller, M.T., & Spradling, A.C. (2007). Male and female *Drosophila* GSCs: two versions of immortality. *Science,* 316: 402-404. doi:10.1126/science.1140861

Fuller, M.T. (1993). Spermatogenesis, In The development of *Drosophila.* (M. Bate, and A. Martinez-Arias, eds.) Cold Harbor Laboratory Press, Cold Spring Harbor, New York, pp 71-147.

Gao, S., Giansanti, M.G., Buttrick, G.J., Ramasubramanyan, S., Auton, A., Gatti, M., & Wakefield, J.G. (2008). Australin: a chromosomal passenger protein required specifically for Drosophila melanogaster male meiosis. *J. Cell Biol.* 180: 521--535. doi:10.1083/jcb.200708072

Giansanti, M.G., Bonaccorsi, S., Williams, B., Williams, E.V., Santolamazza, C., Goldberg, M.L., & Gatti, M. (1998). Cooperative interactions between the central spindle and the contractile ring during *Drosophila* cytokinesis. *Genes Dev.* 12: 396-410. doi:10.1101/gad.12.3.396.

Giansanti, M. G., Belloni, G. & Gatti, M. (2007). Rab11 Is Required for Membrane Trafficking and Actomyosin Ring Constriction in Meiotic Cytokinesis of *Drosophila* Males. *Mol Biol Cell.* 18: 5034-5047. 2007. doi:10.1091/mbc.E07-05-0415

Goldbach, P., Wong, W., Beise, N., Sarpal, R., Trimble, W. S., & Brill, J. A. (2010). Stabilization of the actomyosin ring enables spermatocyte sytokinesis in *Drosophila*. *Mol. Biol. Cell,* 21, 1482-1493. doi: 10.1091/mbc.E09-08-0714.

Goldberg, M.L., K.C. Gunsalus, R.E. Karess, and F. Chang. (1998). Cytokinesis. In *Dynamics of Cell Division.* S.A. Endow and D.M. Glover, editors. Oxford University Press, Oxford. 270–316.

Gonzalez, C., Casal, J., & Ripol, P. 1989. Relationship between chromosome content and nuclear diameter in early spermatids of *Drosophila melanogaster.* Genet. Res.54: 205-212. doi: 10.1017/S0016672300028664.

Hime, G.R., Brill, J.A., & Fuller, M.T. (1996). Assembly of ring canals in the male germ line from structural components of the contractile ring. *J. Cell Sci.***109**: 2779--2788.

Ikeya, T., Galic, M., Belawat, P., Nairz,K., & Hafen, E. 2002. Nutrient-dependent expression of Insulin-like peptides from neuroendocrine cells in the CNS contributes to growth regulation in *Drosophila. Curr biol.,* 12: 1293-1300. doi:10.1016/S0960-9822(02)01043-6

Ichihara, K., Shimizu, H., Taguchi, O., Yamaguchi, M., & Inoue, Y. H. (2007). A *Drosophila* orthologue of Larp protein family is required for multiple processes in male meiosis. *Cell Struct. Funct.* 32: 89-100. doi:10.1247/csf.07027

Inoue, Y.H., Savoian, M. Suzuki, T. Yamamoto, M., & Glover, D.M.. (2004). Mutation in *orbit/mast*, reveal that the central spindle is comprised of two microtubule populations; those that initiate cleavage and those that propagate furrow ingression. *J. Cell Biol.* 116: 1-12. doi: 10.1083/jcb.200402052

Jean Maines, S.W., & Wasserman, S. (1998). Regulation and execution of meiosis in *Drosophila* males. *Curr. Top. Dev. Biol.* 37: 301-332.

Kondylis, V., tot Pannerden, H.E., Herpers, B., Friggi-Grelin, F. & Rabouille, C. (2007) The golgi comprises a paired stack that is separated at G2 by modulation of the actin cytoskeleton through Abi and Scar/WAVE. *Dev. Cell.* 12, 901–915. doi:10.1016/j.devcel.2007.03.008

Kondylis V and Rabouille C. (2009). The Golgi apparatus: Lessons from *Drosophila. FEBS Letters.* 3827-3838. doi:10.1016/j.febslet.2009.09.048

Kim, M., Lee, J.H., Koh, H., Lee, S.Y., Jang, C., Chung, C.J., Sung, J.H., Blenis, J., Chung, J. (2006). Inhibition of ERK-MAP kinase signaling by RSK during *Drosophila* development. EMBO J. 25(13): 3056--3067. doi:10.1038/sj.emboj.7601180

Kiger, A.A., Jones, D.L., Schulz, C., Rogers, M.B., and Fuller, M.T. (2001). Stem cell self-renewal specified by JAK-STAT activation in response to a support cell cue. *Science,* 294: 2542-2545. doi:10.1126/science.1066707

McKee B.D., Habera L, Vrana J.A. (1992). Evidence that intergenic spacer repeats of *Drosophila* melanogaster rRNA genes function as X-Y pairing sites in male meiosis, and a general model for achiasmatic pairing. *Genetics.* 132:529-544.

Mckee B.D., and G. Karpen, *Drosophila* ribosomal RNA genes function as an X-Y meiotic pairing site. *Cell* 61 (1990), pp. 61–72. doi:10.1016/0092-8674(90)90215-Z

Misteli, T. (1997). The mammalian Golgi apparatus during M-phase. *Prog. Cell Cycle Res.* 2: 267, pp.267-277, doi:0.1007/978-1-4615-5873-6_24

Pandey, R. Heidmann, S. Lehner, C.F. (2005). Epithelial re-organization and dynamics of progression through mitosis in Drosophila separase complex mutants. *J. Cell Sci.* 118: pp.733--742

Rabouille, C. Jokitalo, E. (2003). Golgi apparatus partitioning during cell division. *Mol. Membr. Biol.* 20: pp.117-27. doi:10.1080/0968768031000084163

Rabouille, C. & Klumperman, J. (2005). The maturing role of COPI vesicles in intra-Golgi transport. *Nat Rev Mol Cell Biol.* 6: 812-817. doi:10.1038/nrm1735

Rebollo, E., & Gonzalez, C. (2000). Visualising the spindle checkpoint in Drosophila. *EMBO Report.* 1: 65-70. doi:10.1093/embo-reports/kvd011

Rulifson, E.J., Kim, S.K., & Nusse, R. (2002). Ablation of insulin-producing neurons in flies: growth and diabetic phenotypes. *Science,* 296: 1118-1120. doi:10.1126/science.1070058

Rasmussen, S. W. (1973). Ultrastructural studies of spermatogenesis in Drosophila nelanogaster Meigen. Z. Zellforsch. 140, 125-144. doi:10.1007/BF00307062

Savoian, M.S., Goldberg, M.L., & Rieder, C.L. (2000). The rate of poleward chromosome motion is attenuated in Drosophila zw10 and rod mutants. Nat. Cell Biol.2: 948−952. doi:10.1038/35046605

Somma, M. P. Fasulo, B. Cenci, G. Cundari, E. & Gatti, M. (2002). Molecular dissection of cytokinesis by RNA interference in Drosophila cultured cells. *Mol Biol Cell.* 13, pp.2448-2460. doi: 10.1091/mbc.01-12-0589

Soltani-Bejnood, M. Thomas, S. E. Villeneuve, L. Schwartz, K. Hong, C-S. & McKee, B. D. (2007). Role of the mod(mdg4) common region in homolog segregation in Drosophila male meiosis. *Genetics* 176(1): 161−180. doi:10.1534/genetics.106.063289

Shima, D.T., & Warren, G. (1998). Inheritance of the cytoplasm during cell division. In *Dynamics of cell division.* (S.A. Endow, and D.M. Glover eds.) Oxford Press, Oxford, pp 248-269.

Savoian, M.S. Gatt, M.K. Riparbelli, M.G. Callaini, G. & Glover, D.M. (2004). *Drosophila* KLP67A is required for proper chromosome congression and segregation during meiosis I. *J. Cell Sci.* 117: 3669-3677. doi:10.1242/jcs.01213

Tates, A.D. (1971). Cytodifferentiation during spermatogenesis in Drosophila melanogaster. An electron microscope study. Ph.D. thesis, Rijksuniversiteit, Leiden.

Ueishi S, Shimizu H, & Inoue Y. H. (2009). Male germline stem cell division and spermatocyte growth require insulin signaling in Drosophila. *Cell Struct Funct.* 34, 61-9. doi:10.1247/csf.08042

Tomkiel, J.E., Wakimoto, B.T., & Briscoe, A. (2001). The *teflon* gene is required for maintenance of autosomal homolog pairing at meiosis I in male *Drosophila melanogaster. Genetics* 157(1): 273--281.

Vazquez, J., Belmont, A. S., & Sedat, J. W. (2002). The dynamics of homologous chromosome pairing during Male *Drosophila* Meiosis. *Cur. Biol.,* 12, 1473-1483. doi:10.1016/S0960-9822(02)01090-4

White-Cooper, H. (2004). Spermatogenesis, Analysis of meiosis and morphogenesis. *Methods Mol.Biol.* 247: 45-75.

Wilson, P.G., (2005). Centrosome inheritance in the male germ line of Drosophila requires hu-li tai-shao function. *Cell Biol. Int.* 29, 360–369. doi:10.1016/j.cellbi.2005.03.002

Tulina, N., and Matunis, E. (2001). Control of stem cell self-renewal in *Drosophila* spermatogenesis by JAK-STAT signaling. *Science*, 294, 2546-2549. doi:10.1126/science.1066700

Yaffe, M.P., Stuurman, N., and Vale, R.D. (2003). Mitochondrial positioning in fission yeast is driven by association with dynamic microtubules and mitotic spindle poles. *Proc. Natl. Acad. Sci. U. S. A.* 100, 11424-11428.

Yamashita, Y.M. Jones, D. L. Fuller, M. T. (2003) Orientation of asymmetric stem cell division by the APC Tumor Suppressor and Centrosome. Science 301: pp. 1547-1550. DOI: 10.1126/science.1087795

Yamashita, Y.M., Mahowald, A/P. Perlin J.R. & Fuller, M.T. (2007). Asymmetric inheritance of mother versus daughter centrosome in stem cell division. *Science,* 315: 518-521. doi:10.1126/science.1134910

Yamamoto, M. (1979). Cytological studies of heterochromatin function in the *Drosophila melanogaster* male: Autosomal meiotic pairing. Chromosoma 72:293-328. doi:10.1007/BF00331091

Zhou, Q. & Griswold, M.D. (2008) Regulation of spermatogonia, StemBook, ed. The Stem Cell Research Community, StemBook, doi/10.3824/stembook.1.7.1

7

Evaluation of Interspecific-Interploid Hybrids (F₁) and Back Crosses (BC₁) in *Hylocereus* Species (Cactaceae)

Aroldo Cisneros and Noemi Tel-Zur

French Associates Institute for Agriculture and Biotechnology of Drylands,
The Jacob Blaustein Institutes for Desert Research (BIDR),
Ben-Gurion University of the Negev (BGU), Beer Sheva,
Israel

1. Introduction

1.1 Background - *Hylocereus* species

Vine cacti are night-blooming epiphytes plants, endemic to the Americas, and belong to Cactaceae, subfamily Cactoideae, tribe Hylocereeae (Br. and R.) Buxbaum (Barthlott & Hunt, 1993). According to the New Cactus Lexicon (Hunt, 2006), the genera *Hylocereus* (Berger) Br. and R., comprises 14 species, and they are widely distributed in tropical and subtropical regions of the Americas from Mexico to North Argentina (Mizrahi & Nerd, 1999; Merten, 2003). They inhabit a wide range of ecosystems, including coastal areas, high mountains and tropical rainforests (Ortiz, 1999). These species, known as pitahaya or Dragon fruit, are currently being marketed worldwide and have a high economic potential as exotic fruit crops in arid regions where water is scarce, since they use a Crassulacean acid metabolism (CAM) pathway and are exceptionally drought-tolerant (Raveh et al., 1998; Mizrahi & Nerd, 1999; Nobel & de la Barrera, 2004).

Hylocereus species are characterized by triangular stems, bearing large edible fruits (200-600 g) with broad scales and various peels and flesh colours (Lichtenzveig et al., 2000). These species have big flowers (25-30 cm) with large male and female reproductive organs, facilitating manual pollination and manipulation for breeding studies (e.g. the anthers can be easily removed to avoid contamination). The number of ovules per flower is very high, with ~7,200, 5,300 and 2,000 ovules per locule in *Hylocereus undatus* (Haw.) Britton et Rose, *H. monacanthus* (Lem.) Britton et Rose, and *H. megalanthus* (K. Schumann ex Vaupel) Bauer, respectively (Nerd & Mizrahi, 1997; Bauer, 2003; Tel-Zur et al., 2005). *H. megalanthus*, resembles *Hylocereus* in its vegetative appearance, but bears medium-sized (80-200 g) yellow fruit with a spiny peel and a white tasty pulp (Weiss et al., 1994). Seed viability varies among different *Hylocereus* species, as well as within the same species. The pollen donor has an effect on fruit weight, and a positive correlation was found between fruit weight and total seed number when self-pollinated fruit was compared with cross-pollinated fruit (Weiss et al., 1994; Lichtenzveig et al., 2000). *H. monacanthus* was reported to be self-incompatible (Weiss et al., 1994; Lichtenzveig et al., 2000), while *H. undatus,* under growing conditions in Israel, is self-compatible at the end of the season, yet its fruits are smaller than those obtained following cross-pollination. Because of the low yields obtained after self-pollination, hand-cross pollination is routinely carried out for both *H. monacanthus* and *H. undatus* (Mizrahi et al., 2004).

Over the past 20 years, in order to develop this crop to its full economic potential, a traditional breeding program has been developed at Ben-Gurion University of the Negev, focused on the production of superior hybrids in terms of fruit quality, though few of them currently grow at a commercial scale (Tel-Zur et al., 2004, 2005).

2. Polyploidy

Polyploidy is the state of having more than two full sets of homologous chromosomes. Polyploidization may have played a major role in the diversification and speciation of the plant kingdom, generating the genetic and epigenetic novelty that has resulted in the significant diversity observed today (Stebbins, 1971; Soltis & Soltis, 1993). Two forms of polyploidization are autopolyploidization and allopolyploidization. Allopolyploids result from the combination of two genetically and evolutionarily different (or homeologous) genomes. Allopolyploids originated from wide hybridization, as in the case of *Triticum turgidum* L. (Zhang et al., 2010). Enzymes encoded by both parents' alleles and novel enzymes that are encoded by new allelic combinations may be produced by allopolyploids, possibly contributing to their evolutionary success (Soltis & Soltis, 1993). Autopolyploids result from the combination of genomes from two individuals of the same species, producing multiple chromosome sets with similar (homologous) genomes. Autopolyploids can arise from a spontaneous, naturally occurring genome doubling, such as *Galax urceolata* (Soltis et al., 2007).

3. Meiosis and fertility of polyploids

The ploidy level refers to the number of sets of chromosomes (basic number) and is notated by an "x". The basic number is the haploid number of the diploid species, being the chromosome number in a polyploid series, divisible by its basic number (Ranney, 2000).

Both diploid and polyploid organisms can produce viable germ cells. Those of tetraploid organisms are diploid. Since four chromosomes are homologous, quadrivalents are frequently formed during meiosis. Their stability is much less than that of bivalents, leading to an increased ratio of mistakes and, therefore, to reduced fertility and, in extreme cases, even to gamete sterility. Some species have an undisturbed quadrivalent development, while, in others, it does not take place at all (Elliot, 1958).

Polyploid crop species include potatos, coffee, bananas, peanuts, tobacco, wheat, oats, sugarcane, plums, loganberries and strawberries (Stebbins, 1950). Frequently, polyploidization results in bigger organs and improved traits; thus, plant breeders have become interested in the artificial induction of polyploids. It must be pointed out that, in many seed crops, polyploidy lines have shown lower fertility rates than their diploid prototypes; also, in general, there is an optimum range of polyploidy, beyond which growth may be depressed along with increasing chromosome numbers (Elliot, 1958; Wolfe, 2001).

3.1 Meiosis of triploids

Triploids have generally been considered an evolutionary dead end because they have very low fertility and tend to produce aneuploid or unbalanced gametes (Ollitrault et al., 2008). Triploid formation is commonly caused by the fertilization of a haploid egg with diploid pollen or vice versa (Figure 1). During the meiosis of triploids, trivalents are frequently formed. During Anaphase I, the chromosomes are distributed onto both daughter cells. Only in rare cases does one of them receive exactly double the amount ($2n$) of the simple set ($1n$). Generally, both of them are equipped with incomplete sets (aneuploidy; Figure 1). This nearly always results in an imbalance of the chromosome composition, leading to lethality. Therefore, triploidy causes , with a few rare exceptions, sterility of the pollen (or strongly reduced fertility). However, triploids can produce haploid, diploid, or triploid gametes at low rates, which can lead to diploid, triploid, and tetraploid progenies (Otto & Whitton, 2000).

There have been numerous reports in the literature of meiotic behaviour in triploid plants, resulting in valuable cytogenetic information regarding the species investigated. Studies of *Solanum tuberosum* triploids have also provided evidence that the amount of lagging chromosomes is, more or less, proportional to the average number of univalents (Lange & Wagenvoort, 1973). In the case of the wild citrus 'Hong Kong' Kumquat (*Fortunella hindsii* Swing), early cytogenetic studies have described triploid meiosis, showing either trivalent pairing and some univalents (Longley, 1926) or the predominance of trivalents but, also, the presence of numerous bivalents and univalents, as well as great variation in the number of some genotypes (Frost & Soost, 1968).

In general, the triploid plants have failed to set seed with its own pollen; therefore, only in rare cases in citrus and only under controlled conditions was possible to obtain seeds (Ollitrault et al., 2008). In the case of citrus, the seeds obtained from the triploid plants were produced with pollen from diploid species (Ollitrault et al., 2008). Wakana et al. (1981) have shown that the formation of triploid embryos are associated with a pentaploid endosperm, which is a strong indication that triploid hybrids result from the fertilization of unreduced ($2n$) ovules by normal haploid ($1n$) pollen (Esen et al., 1979).

3.2 Meiosis of tetraploids

Polyploidy plays an important role in plant evolution, and it is known that the genomes of flowering plants, including many crop plants of worldwide importance, are polyploidy (Doyle et al., 2008; Leitch & Leitch, 2008).

The meiotic behaviour of plant chromosomes is affected by both genetic and environmental factors (Rezaei et al., 2010). Gametic viability is generally lower in autotetraploid genotypes that have multivalent chromosome association during meiosis than in allotetraploids that form bivalents, leading to equilibrated disomic segregation (Ollitrault et al., 2008). In citrus, it has been shown that the degeneration of pollen mother cells is more frequent in autotetraploids than in their diploid parental genotypes (Frost & Soost, 1968). These authors also observed a great variability in chromosome conjugation (quadrivalents, trivalents, bivalents, and univalents) during Metaphase I and showed that one third to one half of sporads have more than the normal number of four microspores (generally six or seven).

Meiotic restitution (working in functional unreduced female and male gametes) plays a predominant role in producing allopolyploids in flowering plants. This phenomenon, including first division restitution (FDR) and second division restitution (SDR), has been documented in cereal crops (Jauhar, 2003, 2007; Matsuoka & Nasuda, 2004; Zhang et al., 2007, 2010; Wang et al., 2010). It has been suggested that the unreduced gametes produced through meiotic restitution may have been a major mechanism for the widespread occurrence of polyploidy in nature (Jauhar, 2003, 2007; Ramanna & Jacobsen, 2003).

During the meiosis of tetraploids, trivalents and quadrivalents are formed. In Figure 1, the possible combination of balanced and unbalanced gametes in tetraploids during meiosis and in the following Anaphase I and the chromosome distribution onto both daughter cells are shown. Contrary to triploids, tetraploids frequently result with exactly double the amount (2n) of the simple set (1n).

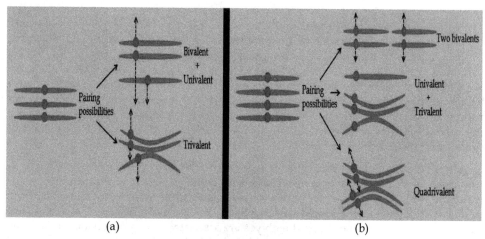

(a)   (b)

Fig. 1. Production of balanced and unbalanced gametes in (a) triploids and (b) tetraploids during meiosis that may to lead balanced and unbalanced segregation and partitioning of chromosome sets.

4. Interspecific-interploid crosses

The following are crucial questions in allopolyploid research: what are the genetic and functional consequences of combining two genetic systems into a common nucleus (Huxley, 1942), and what are their ultimate effects on plant development (Steimer et al., 2004), flowering time (Simpson, 2004) and plant evolution (Kalisz & Purugganan, 2004). The critical period of allopolyploid formation seems to be immediately after fertilization, which probably involves intensive genetic, genomic and physiological changes (McClintock, 1984). In maize, for example, reciprocal interploidy crosses resulted in very small and largely infertile seeds with defective endosperms (Pennington et al., 2008). In *Arabidopsis thaliana*, reciprocal crosses between diploid and tetraploid plants resulted in viable triploid seeds; however, the seed development pattern and, in particular, the endosperm were abnormal (Scott et al., 1998). Additionally, Bushell et al. (2003), working with interspecific-interploid crosses between the diploid *A. thaliana* (2n = 2x = 10) as the seed parent and the tetraploid *A.*

arenosa (2n = 4x = 32) as the pollen donor, produced fruits with aborted seeds, while the reciprocal cross failed to produce fruits, and the flowers collapsed one day after pollination. Several theories have been developed to explain the success or failure of such crosses. Müntzing (1930) postulated that any deviation from the 2:3:2 maternal:endosperm:embryo genome ratios would result in seed failure, but his theory was widely disproved. Von Wangenheim (1957) proposed that seed failure in interploid crosses is caused by shifts in the initial quantity of cytoplasm per nuclear chromosome set, a theory later refuted by Lin (1982). Johnston et al. (1980) put forward the endosperm balance number (EBN) theory, which stated that hybrid success is based on an effective maternal:paternal genome ratio that must be 2:1 for normal endosperm development. This simplified theory was useful for plant breeders for predicting the success of a cross, especially among *Solanum* species (Carputo et al., 1997), but it failed to predict crossing success in other plant species. Haig & Westoby (1989) pointed out that the effect of "parental conflict of interests" on the growth of offspring suggested that maternal and paternal growth-promoting genes would be expressed differentially. Nowack et al. (2006) found that endosperm development is initiated by fertilizing the egg cell only and not the central cell with a mutant of the *A. thaliana* Cdc2 homolog CDC2A (cdc2a mutant pollen produces one sperm cell instead of two), suggesting that a positive signal for proliferation of the central cell triggered endosperm formation following egg fertilization. Nowack et al. (2007) showed that endosperm, exclusively derived from maternal origin, is able to sustain complete embryo development. These findings support Strasburger's (1900) hypothesis that, during evolution, the female gametophyte was reduced to the central cell of modern angiosperms and that fusion of the second sperm is used as a trigger to start endosperm development.

The fertilization process includes the fusion of the gametic membranes and the uptake of sperm cytoplasm by the female gametes, after which the fertilized egg is covered by a cell wall; thus, the endosperm and embryo develop in parallel to form a mature seed.

Crosses between the tetraploid *Hylocereus megalanthus* as the female parent and the diploid *H. undatus* or *H. monacanthus* as the male parent yielded pentaploid, hexaploid and 6x-aneuploid hybrids, while the reciprocal cross, using *H. monacanthus* as the female parent, yielded triploid and 3x-aneuploid hybrids (Tel-Zur et al., 2003). In *H. monacanthus* × *H. megalanthus* crosses, a higher number of truth hybrids were found about 92 %, of them were 3x while the reciprocal cross resulted in 5 % of truth hybrids with a ploidy level higher than expected (Tel-Zur et al., 2003, 2004). Endosperm breakdown is widely believed to be the cause of seed failure; however, seed abortion in vine cactus was probably due to a genomic imbalance between the seed parent and the embryo, rather than the maternal:paternal genome ratio in the endosperm (Tel-Zur et al., 2005) (Table 1). Endosperm dysfunction was the primary reason for hybrid abortion in a range of Angiosperm families (Brink & Cooper, 1947). Nowack et al. (2010) showed that the fertilized egg transmits a signal for development, but it cannot continue without the fertilization of the central cell. On the basis of this hypothesis, Cisneros et al. (2011) postulate, for vine cacti, that: 1) double fertilization happened but the zygote aborted, resulting in empty seeds due to post-zygotic barriers; 2) endosperm formation is necessary (and double fertilization is required) for normal seed coat development, but the presence of an embryo is not essential for the development of a normal black seed coat; and 3) increased genome dosage in the polyploid results in reduced seed viability, which may be attributable to a maternal/paternal imbalance or a lack of double fertilization.

Hybridization is an important source of improved genotypes for cultivation. Hybrids have been produced recently in our laboratory, following interspecific-interploidy crosses among *Hylocereus* species, setting fruits with a range of 5 to 15 % of unviable seed/fruit. The number of the unviable seeds was higher than that observed in the diploid *Hylocereus* species and lower than that observed in the female parent (S–75 or 12-31), suggesting that the triploid and tetraploid level of the hybrids result in a better seed viability. Few of the interspecific-interploidy hybrids set small fruits with a low seed number.

	Seed parent ploidy	Pollen parent ploidy	Predicted seedling ploidy	Predicted endosperm ploidy (m:p)	Seedling ploidy (m:p)	Average seed weight (mg ± SE)
Homoploid crosses	2x	2x	2x	3x (2m:1p)	2x (1m:1p)	2.44 ± 0.05
	4x	4x	4x	6x (4m:2p)	4x (2m:2p)	7.73 ± 0.39
Paternal excess	2x	4x	3x	4x (2m:2p)	3x (1m:2p)	4.37 ± 0.19
Maternal excess	4x	2x	3x	Unreduced female gamete[1]		
				9x (8m:1p) and not 5x (4m:1p)	5x (4m:1p)	3.1 to 5.9
				Chromosome doubling[2]		
				5x (4m:1p)	6x (4m:2p)	3.1 to 5.9
				Aborted seeds[3]		
				5x (4m:1p)	3x (2m:1p)	<3.0

[1]Assuming that all the embryo sac cells (including the egg cell) are a result of unreduced meiosis, since megasporogenesis occurs before megagametogenesis.
[2]Ploidy in the embryo sac cells was assumed as remaining normal, while the hexaploid hybrids are probably a result of chromosome doubling (*see* Tel-Zur et al., 2003).
[3]Triploid hybrids were obtained in this cross direction using embryo rescue (*see* Cisneros, 2009).
(m:p): maternal to paternal genome ratios.
2x: *Hylocereus* species, 4x: *H. megalanthus*.

The results presented in this table are a composite of unpublished and previously published data in Tel-Zur et al. (2005).

Table 1. Outcome of homoploid and interspecific-interploid crosses in species of vine cacti*

5. Phenotypic and genomic evaluation

At the beginning of the 90s, there was very little scientific information available on the cultivation and the biological background of vine cacti. As of now in contrast to the early 90s, investigations in both horticultural and physiological aspects of climbing cacti have been covered and published in the professional literature; beginning with its reproductive biology (Weiss et al., 1994; Nerd & Mizrahi, 1997), shading requirements (Raveh et al., 1998), and fruit development, ripening, and post-harvest handling (Nerd & Mizrahi, 1998; Nerd et al., 1999; Ortiz, 1999).

Phenotypic and genomic characterization of the vine cactus core collection was reported by Tel-Zur et al. (2011), showing a high level of variability for most of the traits studied.

Although the heritability of these traits has yet to be studied, and some are likely to have a substantial environmental component, the levels of variation reported strongly suggested to us that vine cacti have high potential for breeding programs as exotic fruit crops, intrinsically adapted to dry areas (Tel-Zur et al., 2011).

Morphological traits, such as fruit shape, peel and flesh colour, stem size or height, often have one to one correspondence with the genes controlling the traits. In such cases, the morphological characters (the phenotypes) can be used as reliable indicators for specific genes that can be linked with quantitative trait loci (QTL), amplified fragment length polymorphism (AFLP), simple sequence repeat (SSR), among others and must be used to build the genetic linkage map of the vine cacti.

5.1 Phenological characterization

Morphological characteristics have not only been used for formal taxonomical overviews of large germplasm collections, but also for studying geographical patterns, as was done for durum wheat (Yang et al., 1991), bread wheat (Börner et al., 2005), barley (Tolbert et al., 1979) and triticale (Furman et al., 1997). Large variations in morphological characteristics were reported in garlic (Baghalian et al., 2006), different species of onions (Ozodbek et al., 2008; Zammouri et al., 2009), bananas (Uma et al., 2004) and pomegranates (Zamani et al., 2007).

Morphological traits are considered easy to observe, and it is possible to screen and categorize large amounts of genotypes at a low cost, which is a great advantage when managing large germplasm collections (Diederichsen, 2008). However, the optimal utilization of morphological descriptors involves the evaluation of agronomic performance in the farm (Zammouri et al., 2009). The exploitation of such traits increases our knowledge of the genetic variability available and strongly facilitates breeding for wider geographic adaptability, with respect to biotic and abiotic stresses (Diederichsen, 2004).

The vine cacti plants grow up tree trunks and are anchored by aerial roots. The fruits have red, purple or yellow peels, while the colour of the flesh can range from white to red or magenta. The skin is covered with bracts or "scales", hence the name dragon fruit. The seeds are small and are consumed with the fruit. The fruit can weigh up to 900 grams, but the average weight is between 350 and 450 grams (Mizrahi & Nerd, 1999; Merten, 2003). The weight depends on pollination management as well as the genotype (Weiss et al., 1994; Nerd & Mizrahi, 1997). The fruits are most often consumed fresh. In some parts of South America, the pulp is used to prepare drinks (Ortiz, 2001).

One of the major problems in cultivating vine cacti under desert conditions is their sensitivity to both low and high temperatures (Mizrahi & Nerd, 1999). Since variability in these characteristics exist among genotypes, and since there is no genetic barrier among species and even genera, breeding may solve these problems (Lichtenzveig et al., 2000; Tel-Zur, 2001). Another important problem with these fruits that might be solved with breeding is the poor taste of the red pitahayas, *H. monacanthus* and *H. undatus* (Mizrahi et al., 2002). The delicious yellow pitahaya, *H. megalanthus*, can tolerate high temperatures better than the other species, but the fruits are smaller (Mizrahi et al., 2002; Dag & Mizrahi, 2005).

The initial interploid crosses between the diploid *H. monacanthus* as a female parent and the tetraploid *H. megalanthus* as a male parent resulted in the release of two triploid hybrids (named S-75 and 12-31) to growers (Tel-Zur et al., 2004). These hybrids offered better taste than their diploid female parent and larger fruits than their tetraploid male parents, but, still, their spiny peel (a dominant trait), which is characteristic of *H. megalanthus*, was a limitation to further cultivation.

Recently, we performed an evaluation of 109 hybrids for three consecutive years; of this number, 40 were interspecific interploid F_1 (IH) hybrids (among the triploids S-75 or 12-31 and the diploid species *H undatus*), and 69 were first-back crosses (BC$_1$) between the triploids (S-75 and 12-31) and their parental lines. The morphological and agronomical fruit traits from the interspecific interploid F_1 hybrids, as well as the BC$_1$, under study are presented in Table 2, while the characteristics of the parental line were summarized in Tel-Zur et al. (2004, 2005). Variations in fruit form from round to elongated ellipse were observed in the basis of the dissimilarities of the fruit shape index (lower values were shown in hybrids not listed in the Table 2).

The number of flowers per hybrid per year was evaluated in 78 hybrids that set fruits. Among them, a wide range was observed, from one for BC-022 (not listed in Table 2) to 21 for IH-052 (Table 2). The weight of the fruits ranged from 119 to 273 grams and was intermediate between the fruit weights of the parental lines. Comparing these results with the parental genotypes, the majority of the hybrids were intermediate or lower than those of the parents, indicating partial dominance or co-dominance, similar to that described by Tel-Zur et al. (2004).

The potential yield per plant, a very important agricultural trait and a target trait in breeding programs, was calculated on the basis of the number of flowers per year and mean fruit weight (Tel-Zur et al., 2011). Two accessions demonstrated high potential yield, IH-003 and IH-052 with 3.25 and 3.72 kg/plant/year, respectively (Table 2). This trait was probably underestimated and will increase when the plants are older. Tel-Zur et al. (2005) reported the fruit weight in the female parent (hybrids S-75 and 12-31) to be between 192 to 267 grams, depending on the pollen donor used. This trait was intermediate between the parents, indicating partial dominance or co-dominance, as was previously described for others hybrids (Tel-Zur et al., 2004, 2011).

The number of viable seeds/fruit was quantified in several hybrids studied, showing a very high percentage of viable seeds (more than 85%), and their total number did not affect fruit size. In general, the phenotype of the seeds was similar to that described in tetraploids vine cacti (Cisneros et al., 2011); thus, we assume that this is a characteristic inherited from *H. megalanthus*.

The flowering season of the F_1 hybrids and BC$_1$ was from August to November. The time to full ripeness was extremely variable among the studied plants. The results showed variation within hybrids belonging to the same cross and ranged from 52 days in IH-011 (12-31 × *H. undatus*) to 156 days in IH-057 (12-31 × *H. undatus*), while in BC$_1$ (S-75 × *H. monacanthus* or *H. megalanthus*), the variation was lower but more similar to that reported in *H megalanthus*. The ripening time was found to be a genotype-specific trait with considerable variability among *H. megalanthus* accessions (Dag & Mizrahi, 2005). Mizrahi & Nerd (1999); later, Tel-Zur et al. (2004, 2011) reported that flowers of *H. megalanthus* that bloomed in early autumn matured in 90 days while those that bloomed later (November and December) matured in 160 days. Thus, this characteristic had an intermediate behaviour, implying that its inheritance was co-dominant and that the influence of low temperature was minimal.

Plant code[1]	Flowers/ plant/year	Fruit shape index (FL/FW) ± SE	Fruit weight (g) ± SE	Flesh/Peel ratio ± SE	Potential yield/plant (kg)*	Ave. days to ripening
IH-001	8	1.9 ± 0.04	131 ± 18	0.88 ± 0.11	1.05	58
IH-002	5	1.9 ± 0.08	273 ± 33	2.36 ± 0.36	1.36	92
IH-003	16	1.8 ± 0.03	203 ± 15	1.39 ± 0.14	3.25	75
IH-004	12	1.8 ± 0.04	119 ± 10	1.74 ± 0.22	1.43	67
IH-005	6	1.8 ± 0.06	199 ± 27	1.23 ± 0.19	1.19	83
IH-006	12	1.8 ± 0.05	169 ± 14	2.12 ± 0.29	2.03	85
IH-007	8	1.8 ± 0.05	231 ± 25	1.68 ± 0.24	1.85	88
IH-008	9	1.9 ± 0.04	203 ± 26	1.25 ± 0.23	1.83	86
IH-009	13	2.0 ± 0.07	197 ± 17	1.39 ± 0.16	2.56	82
IH-011	11	1.6 ± 0.06	203 ± 10	2.22 ± 0.19	2.23	52
BC-026	10	1.7 ± 0.08	170 ± 25	1.82 ± 0.21	1.70	81
BC-027	7	1.7 ± 0.13	204 ± 26	2.61 ± 0.32	1.43	101
BC-028	12	1.9 ± 0.05	220 ± 17	1.85 ± 0.17	2.64	90
BC-029	10	1.8 ± 0.03	150 ± 14	1.87 ± 0.23	1.50	87
BC-036	8	1.9 ± 0.04	180 ± 16	1.47 ± 0.06	1.44	85
BC-045	11	1.8 ± 0.04	171 ± 12	2.04 ± 0.20	1.88	81
BC-047	10	2.0 ± 0.06	158 ± 24	1.39 ± 0.17	1.58	78
BC-049	8	1.9 ± 0.05	200 ± 21	1.20 ± 0.16	1.60	88
IH-050	16	1.8 ± 0.03	181 ± 24	1.81 ± 0.16	2.90	94
IH-051	12	1.9 ± 0.04	176 ± 9	1.86 ± 0.13	2.12	88
IH-052	21	1.9 ± 0.03	177 ± 13	1.43 ± 0.10	3.72	63
IH-057	8	1.6 ± 0.04	153 ± 32	1.11 ± 0.15	1.23	156
BC-066	7	1.9 ± 0.09	149 ± 29	1.59 ± 0.30	1.04	103
IH-070	10	1.7 ± 0.09	134 ± 19	2.03 ± 0.25	1.34	94
IH-074	6	1.8 ± 0.06	177 ± 24	2.09 ± 0.31	1.06	82
BC-075	9	1.9 ± 0.09	169 ± 23	1.78 ± 0.21	1.52	80
BC-077	6	1.9 ± 0.05	228 ± 28	1.69 ± 0.24	1.37	91
BC-098	8	1.8 ± 0.07	152 ± 17	1.90 ± 0.18	1.22	61
IH-106	6	1.8 ± 0.07	177 ± 19	1.64 ± 0.46	1.06	72
IH-107	7	1.9 ± 0.08	191 ± 19	1.70 ± 0.28	1.33	78

[1] Plant codes refer to IH: Interspecific-interploid cross and BC: Back crosses
* Potential yield per plant was calculated as a number of flowers/year × mean fruit weight

Table 2. F$_1$ and BC$_1$ characterization: flowers per plant, fruit shape index (fruit length/fruit width), fresh/peel ratio, fruit weight, potential yield and average number of days until ripening

5.2 Genome size analyses and ploidy estimation

Estimating genome size and ploidy level in putative hybrids is the first step in evaluating the success of the interploid cross. Two methods are generally used: chromosome count and flow cytometry. Flow cytometry is a technique of genome quantification, initially developed for biomedical research and adapted for genetic plant analysis (Segura et al., 2007), that provides an accurate method to estimate ploidy level by measuring the proportions of cells in the G1, S and G2/M stages of the cell cycle (Doležel et al., 1989). The nuclear phase status is, by convention, indicated using the letter "n". The designations (meiotically) reduced and non-reduced, or haplophasic and diplophasic, are preferable to haploid and diploid, respectively, because their meaning is unambiguous. "n" indicates the meiotically reduced chromosome number, 2n the non-reduced number (Greilhuber et al., 2005; Murray, 2005).

Determining the intra- and inter-specific variation of DNA content is important to prove the success of the hybridization event, as well as for breeding programs and for genetic manipulation (Doležel et al., 1994; Doležel & Bartoš, 2005). These data can also be used to calculate cell cycle times, which are needed in genetic studies, and are useful for analysis of plant growth and development (Loureiro et al., 2007).

There are many examples of correlations between C-value variation between species and cellular parameters, such as the duration of the mitotic and meiotic cell cycle and the sizes of cells. From a taxonomic standpoint, intraspecific C-value variation is probably the most significant indicator that proves the presence of more than one genome combined within a species (Bennett, 1972; Murray, 2005; Záveský et al., 2005). The nuclear replication status (G_1 for non-replicated, S for replicating, G_2 for replicated) leads to DNA content changes expressed in terms of 'C'. For instance, 1C can be the DNA content of a young pollen cell nucleus just after meiosis, and 2C the content of a Telophase root tip nucleus (Moscone et al., 2003; Greilhuber et al., 2005).

Bennett & Leitch (2005) developed a Plant DNA C-values database that currently contains data for 7,058 plant species. It combines the DNA C-values database from the Angiosperm, Gymnosperm, the Pteridophyte, and the Bryophyte, together with the addition of the Algae DNA C-values database. Also included in the database are two Cactaceae species, *Opuntia microdasys* (1C = 2.24 pg) and *Rebutia albiflora* (1C = 1.91 pg), and a number of succulent species from the families Asparagaceae, Bromeliaceae, Crassulaceae, Apocynaceae, Xanthorrhoeaceae and a small number of genera in Asteraceae, in which the 2C-DNA values ranged from 1.11 to 16.85 pg. Segura et al. (2007) reported four different ploidy levels of 23 *Opuntia* species determined by flow cytometry, and the amounts of 2C-DNA ranged from 4.17 pg in *Opuntia incarnadilla* Griffiths to 6.53 pg in *Opuntia heliabravoana* Scheinvar.

A flow cytometric analysis was used to determine chromosome numbers and ploidy in *Consolea* species (Cactaceae). Compared to the base number, the mitotic and meiotic counts indicated hexaploid (2n = 66) and octoploid (2n = 88) species and no diploids, with the 2C-DNA values ranging from 4.88 to 9.50 pg (Negron-Ortiz, 2007). The 2C-DNA content was studied for species of *Hylocereus*, *Selenicereus* and *Epiphyllum*, showing a diploid level for all the species studied, except for two cases of a tetraploid level in *H. megalanthus* and *S. vagans* (Bgek.) Britton et Rose (Tel-Zur et al., 2011). The range of the 2C-DNA content varied from 3.21 pg for *S. grandiflorus* spp. *grandiflorus* (L.) Britton et Rose to 8.77 pg for *H. megalanthus* (Tel-Zur et al., 2011).

2C-DNA content and ploidy estimation was studied in the 109 F_1 and BC_1 of vine cacti using flow cytometry (Table 3). Table 3 shows a random sample chosen from all of them. The 2C-DNA amount in these hybrids ranged from 3.30 pg for IH–051 to 11.67 pg for BC–031 (Table 3), comparable with that reported in different *Hylocereus* and *Selenicereus* species (Tel-Zur et al., 2011) and in other cacti species (Negron-Ortiz, 2007; Segura et al., 2007). These results showed that the ploidy level of the hybrids were diploid, triploid, tetraploid and

Plant code[1]	Nuclear DNA content pg/ 2C ± SD	Genome size 1C (Mbp)	Ploidy analyzed*	Ploidy estimated
IH-001	8.52	4166	N.D	4x
IH-002	6.67	3300	N.D	3x
IH-003	8.42	4044	N.D	4x
IH-004	8.43	4122	N.D	4x
IH-005	6.62	3237	N.D	3x
IH-006	8.22	4068	N.D	4x
IH-011	4.38	2141	N.D	2x
BC-016	8.52	3921	N.D	4x
BC-017	8.66	4234	N.D	4x
BC-018	4.04	1975	N.D	2x
BC-019	6.80	3325	N.D	3x
BC-020	8.08	4469	N.D	4x
BC-021	6.55	3281	N.D	3x
BC-023	11.44	5594	N.D	6x
BC-025	7.67	4528	44	4x
BC-026	6.03	3041	N.D	3x
BC-027	6.20	3022	N.D	3x
BC-028	4.22	2063	N.D	2x
BC-029	3.80	1858	N.D	2x
BC-031	11.67	5706	66	6x
BC-032	8.51	4811	N.D	4x
BC-033	5.72	2934	N.D	3x
BC-034	8.00	3735	N.D	4x
BC-035	6.22	2997	33	3x
BC-036	7.25	3545	N.D	4x
BC-037	6.98	3413	N.D	3x
BC-045	6.87	3359	29 – 33	3x or mix
IH-050	3.40	1662	N.D	2x
IH-051	3.30	1613	N.D	2x
IH-052	4.53	2215	N.D	2x
IH-057	9.00	4401	N.D	4x
IH-062	4.33	2117	N.D	2x
IH-106	4.57	2234	N.D	2x

[1] Plant codes refer to IC: Interspecific-interploid cross and BC: Back crosses
* Ploidy analyzed was based on the number of chromosomes counted using acetocarmine stain

Table 3. Genome size and ploidy estimation in F_1 and BC_1.

hexaploid, with only one exception, BC–045, that was triploid or aneuploid (Table 3). In different species of *Hylocereus* and *Selenicereus*, diploid and tetraploid were reported (Tel-Zur et al., 2011). In the BC_1, we expected to find ploidy levels ranging from 2x to 4x, according to the ploidy of the parents, [for example, the cross between the female triploid S–75 or 12–31 (2n=3x=33) and the diploid male parent *H. monacanthus* (2n=2x=22) or the tetraploid *H. megalanthus* (2n=4x=44)], but never 6x, as was obtained for the BC–023 (12–31 × *H. monacanthus*) and BC–031 (S–75 × *H. monacanthus*). In these cases, we assumed that the genome was duplicated following a fertilization event with an unreduced (2n) female gamete from the triploid female parent (S–75 or 12–31) and a normal reduced (n) male gamete from the diploid *H. monacanthus*. Cisneros (2009) reported analogous unexpectedly high ploidy levels (6x) obtained in similar interspecific-interploid crosses between the triploid S–75 and the tetraploid *H. megalanthus* accession 96-667. Ploidy levels higher than 3x, observed in the hybrids under study, were probably due to unreduced gametes produced by the mother plant or by the pollen donor.

6. Cytology of chromosome non-disjunction

One of the major routes for polyploidization involves gametic "non-reduction" or "meiotic nuclear restitution" during microsporogenesis and megasporogenesis, resulting in unreduced 2n gametes. Non-reduction could be due to meiotic non-disjunction, failure of cell wall formation or formation of gametes by mitosis instead of meiosis (Elliot, 1958; Grant, 1981). Unreduced gametes (2n) are recognized as a common mechanism of origin of most polyploids in plants (Sang et al., 2004; Otto, 2007; Matthew et al., 2009). Generally, 2n gametes originate due to deviating meiosis in plants. Deviations can occur in plants with normal chromosome pairing, as well as in those with disturbed chromosome pairing as, for example, in distant interspecific hybrids or synaptic mutants. The process that leads to the formation of 2n gametes is called meiotic nuclear restitution and occurs either during micro- or megasporogenesis (Ramanna & Jacobsen, 2003).

Cytological disturbances may lead to sterility or reduced seed viability. Cytological disturbances and anomalous behaviours, such as heavy-walled coenocytes, uncoiled chromosomes, supernumerary chromosomes, production of cross-bridges at second division and the occurrence of globular structures in the microsporocytes, were reported for several species and are strongly associated with a high level of sterility in polyploids, e.g., in the triploid hybrid of *Gossypium hirsutum-herbaceum* (Beasley, 1940), in the allohexaploid of *Phleum pratense* (Nath & Nielsen, 1961) and in the tetraploids of *Brachiaria decumbens* (Simioni & Borges do Valle, 2011).

Winge (1917) proposed a theory of "hybridization followed by chromosome doubling" as a mechanism enabling the survival and development of the hybrid zygote by providing each chromosome with a homologue with which to pair. Despite the lack of well-documented evidence, generations of biologists believed that polyploids were generated by somatic doubling (zygotic or meristematic). Conversely, Harlan & de Wet (1975) listed 85 plant genera known to produce 2n or "unreduced" gametes (pollen or egg cells carrying the somatic chromosomal numbers), which reinforced the theory of Karpechenko (1927) and of Darlington (1956) that sexual polyploidization, resulting from 2n gamete fusion, is the driving force behind the origin of polyploid species. Currently, most of the evidence attributes polyploid formation to: (1) sexual polyploidization through fusion of 2n gametes; (2) somatic mutations in meristematic cells, namely, chimera (Morgan et al., 2001; Karle et al., 2002); (3) somatic polyploidization by nuclear fusion (Baroux et al., 2004); or (4) polyspermy, the fertilization of an egg by more than one sperm (Virfusson, 1970).

There is, however, some scientific evidence supporting the occurrence of "hybridization followed by chromosome doubling", the best known example being that of *Primula "kewensis"*, a first-generation hybrid between *P. floribunda* and *P. verticillata* (Newton &

Pellew, 1929). For some unknown reason, the chromosomes in one branch of the plant had spontaneously doubled (somatic doubling), a process that apparently restored fertility by providing each chromosome with an identical partner with which to pair. Other examples are the spontaneous appearance of the tetraploid *Oenothera lamarckiana* Ser. (Gates, 1924) and the amphidiploids *Nicotiana glutinosa* L. and *N. tabacum* L. (Clausen & Goodspeed, 1925), which provide empirical evidence for polyploidization, presumably, by chromosome doubling. Furthermore, corn plants exposed to heat shock after pollination produced diploid, tetraploid and octaploid seedlings. The latter two states of polyploidy seem to be the result of somatic doubling in the zygote or young embryo (Müntzing, 1933). Nath & Nielsen (1961) reported that the origin of the hexaploid level in *Phleum pratense* was due to the trebling of the diploid genome complement from the *P. nodosum* species.

Lately, evidence supporting Winge's theory has been reported by Tel-Zur et al. (2003) in vine cacti crosses between the tetraploid *Hylocereus megalanthus* as the female parent and the diploid *H. undatus* or *H. monacanthus* as the male parent, yielding several hybrids with an unexpectedly high ploidy level, always higher than that of the female parent. Since unreduced gametes were not observed in the diploid *Hylocereus* species (Tel-Zur et al., 2003), the origin of 6x is still not clear, and could be the result of chromosome doubling following interspecific-interploid crosses. Cisneros (2009) found similar results following interspecific-interploid crosses by using the embryo rescue technique, in which only one hybrid out of 22 tested showed the expected (3x) ploidy level that was lower than the ploidy of the female parent.

Univalents and multivalents were observed in the pollen mother cells (PMCs) of the tetraploid *H. megalanthus* at Metaphase I, even though chromosome disjunction at Anaphase I was very balanced (Lichtenzveig et al., 2000). The large pollen grains observed in this species, about 12% of the sample, were probably unreduced gametes formed due to meiotic irregularities during Anaphase II (Tel-Zur et al., 2003). Therefore, the pentaploid hybrids reported were probably the result of a fertilization event between an unreduced egg cell (from the tetraploid *H. megalanthus*) and a reduced pollen grain (from the diploid *H. undatus* or *H. monacanthus*). However, the hexaploid and 6x-aneuploid hybrids are an exceptional case, since the diploid *H. undatus* and *H. monacanthus* showed a regular chromosome disjunction at Anaphase I and a uniform pollen diameter, and all the interspecific-homoploid (2x) *Hylocereus* × *Hylocereus* hybrids were diploids, which strongly indicated insignificant or null production of unreduced gametes (Lichtenzveig et al., 2000; Tel-Zur et al., 2003, 2004).

Consequently, the hexaploid hybrids obtained as a result of the interspecific-interploid cross occurred at a frequency much higher than that expected for a possible fusion of 2n gametes from both egg and pollen donor parents. In interspecific-interploidy triploid and 3x-aneuploid hybrids, rod and ring bivalents were observed in the PMCs at Metaphase I. The frequency of the ring bivalents was much lower than that of the rod bivalents, followed by a balanced segregation in Anaphase I (Tel-Zur et al., 2005). Abnormal spindle geometry during Metaphase I (parallel and tripolar) and lagging chromosomes were also observed (Figure 2). Therefore, the relatively high percentages of functional female and male gametes (9.8 - 18.6% of viable pollen and 6.0 - 35.5% of viable seeds) produced by 3x hybrids are most likely the result of balanced chromosome segregation during meiosis (Tel-Zur et al., 2005). Moreover, all the hybrids were fertile or partially fertile, indicating that pre- and post-zygotic barriers are not a major factor blocking the development and viability of hybrids among vine cacti species (Tel-Zur et al., 2003, 2004, 2005).

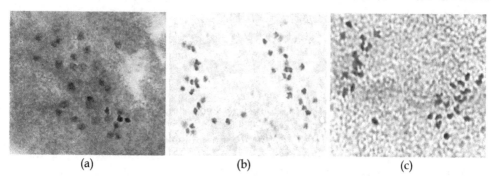

(a)   (b)   (c)

Fig. 2. Meiotic abnormalities observed in interspecific-interploid hybrids. (a) Hybrid BC–045 (3x aneuploid): Prophase I showing 32 chromosomes. (b) Hybrid BC–025 (4x): later Anaphase I showing balanced chromosome disjunction and lagging chromosomes. (c) Hybrid BC–035 (3x): later Anaphase I showing unbalanced chromosome disjunction and a lagging chromosome.

7. Cytomixis

Cytomixis was defined as the migration of chromatin between adjacent cells through a cytoplasmic connection channel, i.e., chromatin migration between meiocytes, and was reported for the first time in the PMCs of *Crocus sativus* by Körnicke in 1901 (as cited in Bellucci et al., 2003). Cytomixis has been extensively studied in a great number of species; however, its origin and significance are still unclear, and its role in evolution, as well as its genetic control, remains speculative and controversial. Cytomixis is now considered to be a cytological phenomenon, though infrequent, and not an artefact that occurred during slide preparations (Bellucci et al., 2003). This phenomenon was observed more frequently during microsporogenesis, mostly during Prophase of the meiotic division, but can occur in all stages of the meiosis, especially in genetically unbalanced genotypes such as haploids, triploids, aneuploids, mutants and hybrids (de Nettancourt & Grant, 1964; Gottschalk, 1970; Salesses, 1970; Mantu & Sharma, 1983). Usually, a few cells (two to four) participated in the process, while a large numbers of PMCs were involved (Malallah & Attia, 2003).

The process of chromatin transference occurs mainly from the donor to the recipient cell and could include a small part of the chromatin material or the whole genome of the donor cell. Negron-Ortiz (2007) reported, in *Consolea* species, the occurrence of cytomixis, and the number of cells involved varied from two to nine, depending on the species, resulting, occasionally, in the establishment of empty microsporocytes after a total chromatin migration. The meiocytes with no chromatin were lost during the meiotic division, and those with abnormal genome size formed unbalanced gametes.

Previous reports indicated a genetic control of the cytomixis process (Omara, 1976; Morikawa & Legget, 1996), which can be affected by extremely high temperatures (Mantu & Sharma, 1983), herbicides (Bobak & Herich, 1978), or by pathological agents (Bell, 1964).

During observations of immature anthers of the F_1 and BC_1 vine cacti, multiple chromatin bridges between microsporocytes were observed at Prophase I; such bridges allow chromatin transfer between cells (Figure 3). These observations of PMCs showed that the cytomixis process detected during the Prophase of the meiosis was more frequent in some of the interspecific interploid hybrids between the allotriploid S–75 as the mother parent, and the diploid *H. undatus*, as the pollen donor. Most likely, this phenomenon took place in these hybrids because of the convergence of three different genomes (*H. monacanthus*, *H. megalanthus* and *H. undatus*), and may thus imply selective DNA elimination as a response to the allopolyploidization process. The number of cells involved in the phenomenon varied from two to four. Completely empty microsporocytes were not observed. In some PMCs' donor, the remark of some chromosomes indicated that the migration to its attached recipient cell was not complete (Figure 3). Another interesting phenomenon observed was the formation of vesicle-like objects in the walls of the PMCs (Figure 3). Chromosome segments were observed in some of those vesicles and around them, which seems to be a way to remove DNA from the PMCs before the meiotic division. To the best of our knowledge, no previous reports were found in the literature addressing similar vesicle-like formation.

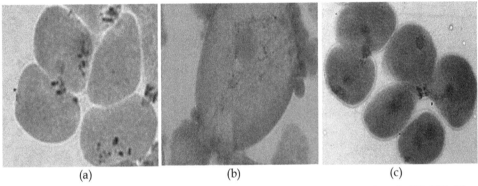

(a)   (b)   (c)

Fig. 3. Cytomixis process in PMCs in vine cacti. (a) Cytoplasmic connections in IH–006. (b) Vesicle-like formation with DNA material during Prophase in BC–023. (c) Prophase showing chromosomes that were not involved in the migration process in BC–032.

The ability of cytoplasmic channels to penetrate the callus walls, which are usually formed around microsporocytes at the end of Prophase, determines the beginning of the cytomixis process. Still unknown are the trigger that induces the PMCs to develop a connection channel with its neighbour; what points are involved in the process of recognizing the potential receptor cell; and if this process is a result of the genetic instability and/or incompatibility due to the combination of different genomes in a single cell.

8. Conclusions

Classical breeding methods can be defined as a set of tools that allows researchers to improve and characterize plants or living organisms. Several technologies, such as fluorescent activated cell sorting (FACS), molecular markers [e.g., random amplified polymorphism (RAP), amplified fragment length polymorphism (AFLP) and fluorescent *in situ* hybridization (FISH)], marker assistant selection (MAS), and quantitative trait loci (QTL) are currently extensively used for the purpose of plant evaluation and characterization. Combining these technologies with phenotypic, genomic and cytological evaluations has given us a model system by which to associate the agronomical trait with the gene controlling its expression and making it possible to infer the inheritance. Additionally, this model system provides an excellent research tool to elucidate the pathways of polyploid formation and seed development following interspecific-interploid hybridization and to solve critical hypotheses related to the origin of vine cacti species, contributing to the new discoveries of improved nutritional value, potential yield enhancement, breaking the self-incompatibility of the diploid lines and improving resistance to abiotic factors.

Two major basic mechanisms are most likely involved in polyploidization in vine cacti: (1) unreduced (2n) gametes, mostly arising from the nuclear division restitution during the meiosis of the micro- or megasporegenesis that can produce hybrids with a ploidy level higher than that of the parents in triploid × diploid and triploid × tetraploid hybridizations; and (2) duplication of the chromosomes after hybridization that gives rise to polyploidy.

In general, the meiocytes resulting from the meiosis in triploids and tetraploids can display more than one possible pairing association (*see* Figure 1); thus, the probability to produce unreduced (2n, 3n or 4n) and unbalanced gametes is high.

The results summarized here provide experimental support to the hypothesis of polyploidization through somatic chromosome doubling or due to meiotic non-disjunction, although both the timing of the polyploidization event and the nature of the trigger remain unclear, as does the entire process leading to genome doubling. Further work on this topic will be directed to elucidate this phenomenon.

9. Acknowledgments

The authors thank Zin Scholarship Kreitman School and A. Katz International School for Desert Studies (BGU) for partial support of this work with a doctoral fellowship granted to A. Cisneros. This research was supported by Research Grant No. IS-4017-07 from BARD, the United States–Israel Binational Agricultural Research and Development Fund. We extend our gratitude to Mr. J. Mouyal and to the late Dr. B.Schneider for valuable assistance.

10. References

Baghalian, K., Sanei, M.R., Naghavi, M.R., Khalighi, A. & Naghdi, B.H. (2006). Post-culture evaluation of morphological divergence in Iranian garlic ecotypes. *Scientia Horticulturae,* 107: 405–410.

Baroux, C., Pecinka, A., Fuchs, J., Schubert, I. & Grossniklaus, U. (2007). The triploid endosperm genome of *Arabidopsis* adopts a peculiar, Parental-Dosage-Dependent chromatin organization. *The Plant Cell,* 19: 1782–1794.

Barthlott W, Hunt DR (1993) Cactaceae. In: Kubitzki K (ed) *The families and the genera of vascular plants,* Vol. 2, Springer, Berlin, pp. 161–196.

Bauer, R. (2003). A synopsis of the tribe Hylocereeae. F. Buxb. Cactaceae System. *Initiatives* 17: 3–63.

Beasley, J.O. (1940). The production of polyploids in *Gossypium*. Journal of Heredity, 30: 39–48.

Bellucci, M., Roscini, C. & Mariani, A. (2003). Cytomixis in Pollen Mother Cells of *Medicago sativa* L.. Journal of Heredity, 94: 512–516.

Bennett, M.D. (1972). Nuclear DNA content and minimum generation time in herbaceous plants. *Proc. R. Soc. Lond. B.*, 181: 109–135.

Bennett, M.D. & Leitch, I.J. (2005). Plant DNA C-values database, In: Release 5.0, December 2010, Available from http://www.rbgkew.org.uk/cval/homepage.html

Bobak, M. & Herich, R. (1978). Cytomixis as a manifestation of phatological changes after the application of Trifuraline. *Nucleus*, 21: 22–26.

Börner, A., Schäfer, M., Schmidt, A., Grau, M. & Vorwald, J. (2005). Associations between geographical origin and morphological characters in bread wheat (*Triticum aestivum* L.). *Plant Genetic Resource*, 3: 360–372.

Brink, R.A. & Cooper, D.C. (1947). The endosperm in seed development. *The Botanical Review*, 13: 423–541.

Bushell, C., Spielman, M. & Scott, R.J. (2003). The basis of natural and artificial postzygotic hybridization barriers in *Arabidopsis* species. *Plant Cell*, 15: 1430–1442.

Carputo, D., Barone, A., Cardi, T., Sebastiano, A., Frusciante, L. & Peloquin, S.J. (1997). Endosperm balance number manipulation for direct *in vivo* germplasm introgression to potato from a sexually isolated relative (*Solanum commersonii* Dun.). *Proceeding of National Academy of Science*, USA, 94: 12013–12017.

Cisneros, A. (2009). Seed development following interespecific-homoploid and interespecific-interploid crosses in species of vine cactus. In: *Thesis submitted in partial fulfilment of the requirements for the degree of Master of Science*, (April 2009), pp. 9–46, Ben-Gurion University of the Negev, Beer Sheva, Israel.

Cisneros, A., Benega Garcia, R. & Tel-Zur, N. (2011). Ovule morphology, embryogenesis and seed development in three *Hylocereus* species (Cactaceae). *Flora*206 (12): 1076– 1084, ISSN 0367-2530.

Clausen, R. E. & Goodspeed, T.H. (1925). Interspecific hybridization in *Nicotiana*. II. A tetraploid *glutinosa-tabacum* hybrid, an experimental verification of Winge's hypothesis. *Genetics*, 10: 278–284.

Dag, A. & Mizrahi, Y. (2005). Effect of pollination method on fruit set and fruit characteristics in the vine cactus Selenicereus megalanthus ("yellow pitaya"). J *Hortic. Sci. Biotechnol.*, 80: 618–622.

Darlington, C.D. (1956). *Chromosome botany*, Allen & Unwin, London, England.

de Nettancourt, D. & Grant, W.F. (1964). La cytogenetique de *Lotus (leguminosae)* III. Un cas de cytomixie dans un hybride interspecifique. *Cytologia*, 29: 191–195.

Diederichsen, A. (2004). Case studies for the use of infraspecific classifications in managing germplasm collections of cultivated plants. *Acta Horticulturae*, 634: 127–139.

Diederichsen, A. (2008). Assessments of genetic diversity within a world collection of cultivated hexaploid oat (*Avena sativa* L.) based on qualitative morphological characters. *Genetic Resources and Crop Evolution*, 55: 419–440.

Doležel, J. & Bartoš, J. (2005). Plant DNA flow cytometry and estimation of nuclear genome size. *Annal of Botany*, 95: 99–110.

Doležel, J., Binarová, P. & Lucretti, S. (1989). Analysis of nuclear DNA content in plant cells by flow cytometry. *Biol. Plant.*, 31: 113–120.

Doležel, J., Doleželová, M. & Novák, F.J. (1994), Flow cytometric estimation of nuclear DNA amount in diploid bananas (*Musa acuminata* and *M. balbisiana*). *Biol. Plant.*, 36: 351-357.

Doyle, J.J., Flagel, L.E., Paterson, A.H., Rapp, R.A., Soltis, D.E., Soltis, P.S. & Wendel J.F. (2008). Evolutionary genetics of genome merger and doubling in plants. *Annu. Rev. Genet.*, 42: 443-461.

Elliot, F.C. (1958). *Plant breeding and cytogenetics*. McGraw Hill Book Co., pp. 10-178, New York, USA.

Esen, A., Soost, R. K. & Geraci, G. (1979). Genetic evidence for the origin of diploid megagametophytes in Citrus. *Journal of Heredity*, 70: 5-8.

Frost, H.B. & Soost, R.K. (1968). Seed reproduction: Development of gametes and embryos. In: *The citrus industry*, W. Reuther, L.D. Batchelor, and H.B. Webber (eds.), Univ. California Press, Vol. 2, pp. 290-324.

Furman, B.J., Qualset, C.O., Skovmand, B., Heaton, J.H., Corke, H. & Wesenberg, D.M. (1997). Characterization and analysis of North American triticale genetic resources. *Crop Science*, 37: 1951-1959.

Gates, R. R. (1924). Polyploidy. *Brit. J. Exp. Biol.*, 1: 153-182.

Gottschalk, W. (1970). Chromosome and nucleus migration during microsporogenesis of *Pisum sativum*. *Nucleus*, 13: 1-9.

Grant, V.P. (1981). *Polyploidy*, Columbia University Press, pp. 283-352, New York, USA.

Greilhuber, J., Doležel, J., Lysák, M.A. & Bennett, M.D. (2005). The origin, evolution and proposed stabilization of the terms 'Genome Size' and 'C-Value' to describe nuclear DNA contents. *Annal of Botany*, 95: 255-260.

Haig, D. & Westoby, M. (1989). Parent-specific gene-expression and the triploid endosperm. *Am. Nat.*, 134: 147-155.

Harlan, J.R. & de Wet, J.M.J. (1975). On Ö. Winge and a prayer: The origins of polyploidy. *The Botanical Review*, 41: 361-390.

Hunt, D. (2006). Hunt, D. (2006). *The New Cactus Lexicon*. Remous Ltd., ISBN 0953813444, Milborne Port, England.

Huxley, J. (1942). *Evolution: the modern synthesis*, Harper and Brothers, London, England.

Jauhar, P.P. (2003). Formation of 2n gametes in durum wheat haploids: sexual polyploidization. *Euphytica*, 133: 81-94.

Jauhar, P.P. (2007). Meiotic restitution in wheat polyhaploid (amphihaploids): a potent evolutionary force. *Journal of Heredity*, 98: 188-193.

Johnston, S.A., den Nijs, T.P.M., Peloquin, S.J. & Hanneman, R.E. (1980). The significance of genetic balance to endosperm development in interspecific crosses. *Theoretical and Applied Genetics*, 57: 5-9

Kalisz, S.M. & Purugganan, M.D. (2004). Epialleles via DNA methylation: consequences for plant evolution. *Trends in Ecology and Evolution*, 19: 309-314.

Karle, R., Parks, C.A., O'Leary, M.C. & Boyle, T.H. (2002). Polyploidy-induced changes in the breeding behavior of *Hatiora xgraeseri* (Cactaceae). *J. Am. Soc. Hort. Sci.*, 127: 397-403.

Karpechenko, G.D. (1927). The production of polyploid gametes in hybrids. *Hereditas*, 9: 349-368.

Lange, W. & Wagenvoort, M. (1973). Meiosis in triploid *Solanum tuberosum* L. *Euphytica*, 22: 8-18.

Leitch, A.R. & Leitch, I.J. (2008). Genomic plasticity and the diversity of polyploid plants. *Science*, 320: 481–483.

Lichtenzveig, J., Abbo, S., Nerd, A., Tel-Zur, N. & Mizrahi, Y. (2000). Cytology and mating systems in the climbing cacti Hylocereus and Selenicereus. *American Journal of Botany*, 87: 1058–1065.

Lin, B.Y. (1982). Association of endosperm reduction with parental imprinting in maize. *Genetics*, 100: 475–486.

Longley, A.E. (1926). Triploid *Citrus*. *J. Wash. Acad. Sci.*, 16: 543–545.

Loureiro, J., Rodriguez, E., Doležel, J. & Santos, C. (2007). Two new nuclear isolation buffers for plant DNA flow cytometry: A test with 37 species. *Annal of Botany*, 100: 875–888.

Malallah, G.A. & Attia, T.A. (2003). Cytomixis and its possible evolutionary role in a Kuwaiti population of *Diplotaxis harra* (Brassicaceae). *Botanical Journal of the Linnean Society* 143: 169–175.

Mantu, D.E. & Sharma, A.K. (1983). Cytomixis in pollen mother cell of an apomitic ornamental *Ervatamia divaricata* (Linn) Alston. *Cytologia* 48: 201–207.

Matsuoka, Y. & Nasuda, D.S. (2004). Durum wheat as a candidate for the unknown female progenitor of bread wheat: an empirical study with a highly fertile F1 hybrid with *Aegilops tauschii* Coss. *Theoretical and Applied Genetics*, 109: 1710–1717.

Matthew, N.N., Mason, A.S., Castello, M.C., Thomson, L., Yan, G. & Cowling, W.A. (2009). Microspore culture preferentially selects unreduced (2n) gametes from an interspecific hybrid of *Brassica napus* L. × *Brassica carinata* Braun. *Theoretical and Applied Genetics*, 119: 497–505.

McClintock, B. (1984). The significance of responses of the genome to challenge. *Science*, 226: 792–801.

Merten, S. (2003). A review of *Hylocereus* production in the United States. *J PACD*, 5: 98–105.

Mizrahi, Y. & Nerd, A. (1999). Climbing and columnar cacti: New arid land fruit crops. In: *Perspective in new crops and new uses*, J. Janick (ed.), ASHS Press, Alexandria, VA, pp: 358–366.

Mizrahi, Y., Nerd, A. & Sitrit, Y. (2002). New fruits for arid climates. In: *Trends in new crops and new uses*, J. Janick and A. Whipkey (eds.), ASHS Press, Alexandria, VA, pp: 378–384.

Mizrahi, Y., Mouyal, J., Nerd, A. & Sitrit, Y. (2004). Metaxenia in the vine cacti *Hylocereus polyrhizus* and *Selenicereus* spp. *Annal of Botany*, 93: 469–472.

Morgan, W.G., King, I.P., Koch, S., Harper, J.A. & Thomas, H.M. (2001). Introgression of chromosomes of *Festuca arundinacea* var. glaucescens into *Lolium multiflorum* revealed by genomic in situ hybridisation (GISH). *Theoretical and Applied Genetics*, 103: 696–701.

Morikawa, T. & Leggett, J.M. (1996). Cytological variations in wild populations of *Avena canariensis* from the Canary Islands. *Genes Genet. Syst.*, 71: 15–21.

Morisset, P. (1978). Cytomixis in the pollen mother cells of Ononis (Leguminosae). *Can. J. Genet. Cytol.*, 20: 383–388.

Moscone, E.A., Baranyi, M., Ebert, I., Greilhuber, J., Ehrendorfer, F. & Hunziker, A.T. (2003). Analysis of nuclear DNA content in Capsicum (Solanaceae) by flow cytometry and feulgen densitometry. *Annal of Botany*, 92: 21–29.

Müntzing, A. (1930) Uber chromosomenvermehrung in *Galeopsis* – Kreuzungen und ihre phylogenetische Bedeutung. *Hereditas*, 14: 153–172.

Müntzing, A. (1933). Hybrid incompatibility and the origin of polyploidy. *Hereditas*, 18 (1-2): 33–55.

Murray, B.G. (2005). When does intraspecific C-value variation become taxonomically significant?, *Annal of Botany*, 95: 119–125.

Nath, J. & Nielsen, E.L. (1961). Cytology of plants from self and open-pollination of *Phleum pratense*. *American Journal of Botany*, 48 (9): 772–777.

Negron-Ortiz, V. (2007). Chromosome numbers, nuclear DNA content, and polyploidy in Consolea (Cactaceae), an endemic cactus of the Caribbean islands. *American Journal of Botany*, 94: 1360–1370

Nerd, A. & Mizrahi, Y. (1997). Reproductive biology of cactus fruit crops. *Hort. Rev.*, 18: 321–346.

Nerd, A. & Mizrahi, Y. (1998). Fruit development and ripening in yellow pitaya. *J. Amer. Soc. Hortic. Sci.*, 123: 560–562.

Nerd, A., Guttman, F. & Mizrahi, Y. (1999). Ripening and postharvest behaviour of fruits of two *Hylocereus* species (Cactaceae). *Postharvest Biol. Technol.*, 17: 39–45.

Newton, W.C.F. & Pellew, C. (1929). *Primula kewensis* and its derivatives. *Journal of Genetics*, 20 (3): 405–467.

Nobel, P.S. & de la Barrera, E. (2004). CO_2 uptake by the cultivated hemiepiphytic cactus, *Hylocereus undatus*. *Ann. Appl. Biol.*, 144: 1–8.

Nowack, M.K., Grini, P.E., Jakoby, M.J., Lafos, M., Koncz, C. & Schnittger, A. (2006). A positive signal from the fertilization of the egg cell sets of endosperm proliferation in angiosperm embryogenesis. *Nature Genet.*, 38: 63–67.

Nowack, M.K., Shirzadi, R., Dissmeyer, N., Dolf, A., Endl, E., Grini, P.E. & Schnittger, A. (2007). Bypassing genomic imprinting allows seed development. *Nature*, 447: 312–316.

Nowack, M.K., Ungru, A., Bjerkan, K.N, Grini, P.E. & Schnittger, A. (2010). Reproductive cross-talk, seed development in flowering plants. *Bioch. Soc. Trans.*, 38: 604–612.

Ollitrault, P., Dambier, D., Luro, F. & Froelicher, Y. (2008). Ploidy Manipulation for Breeding Seedless Triploid Citrus. In: *Plant Breeding Reviews*, J. Janick (ed), Vol. 30, pp: 323–353, John Wiley & Sons Inc., Hoboken, NJ, USA. doi:10.1002/9780470380130.ch7

Omara, M.K. (1976). Cytomixis in *Lolium perenne*. *Chromosoma*, 55: 267–271.

Ortiz, Y.D.H. (1999). Pitahaya a new crop for Mexico. Editorial Limusa, S.A. de C.V. Grupo Noriega Edit., pp: 35–58, ISBN 968-18-5775-5.

Ortiz, Y.D.H. (2001). Hacia el conocimiento y conservacion de la pitahaya (*Hylocereus* spp.). In: *IPN-SIBEJ-CONACYT-FMCN*, Oaxaca, Mexico.

Otto, S.P. & Whitton, J. (2000). Polyploid incidence and evolution. *Annu. Rev. Genet.*, 34: 401–437.

Otto, S.P. (2007). The evolutionary consequences of polyploidy. *Cell*, 131: 452–462.

Ozodbek, A.A., Svetlana, S.Y. & Reinhard, M.F. (2008). Morphological and embryological characters of three middle Asian *Allium* L. species (Alliaceae). *Bot. J. Lin. Soc.*, 137 (1): 51–64.

Pennington, P.D., Costa, L.M., Gutierrez-Marcos, J.F., Greenland, A.J. & Dickinson, H.G. (2008). When genomes collide: Aberrant seed development following maize interploidy crosses. *Annal of Botany*, 101: 833–843.

Ramanna, M.S. & Jacobsen, E. (2003). Relevance of sexual polyploidization for crop improvement–a review. *Euphytica*, 133: 3–18.

Ranney, T.G. (2000). Polyploidy: From Evolution to Landscape Plant Improvement. *Proceedings of the 11th Metropolitan Tree Improvement Alliance (METRIA)*. Gresham - Oregon, USA, August 23–24, 2000.

Raveh, E., Nerd, A. & Mizrahi, Y, (1998). Responses of two hemiepiphytic fruit-crop cacti to different degrees of shade. *Scientia Horticulturae*, 73 (2, 3): 151–164.

Rezaei, M., Arzani, A. & Sayed-Tabatabaei, B.E. (2010). Meiotic behaviour of tetraploid wheats (*Triticum turgidum* L.) and their synthetic hexaploid wheat derivates influenced by meiotic restitution and heat stress. *J. Genet.*, 89: 401–407.

Salesses, G. (1970). Sur le phenomene d cytomixie chez des hybrids triploides de prunier. Consequences genetiques possibles. *Annales del amerioration des plantes*, 20: 383–440.

Sang, T., Pan, J., Zhang, D., Ferguson, D., Wang, C., Hong, D.Y. & Pan, K.U. (2004). Origins of polyploids: an example from peonies (Paeonia) and a model for angiosperms. *Biol. J. Linn. Soc.*, 82: 561–571.

Scott, R.J., Spielman, M., Bailey, J. & Dickinson, H.G. (1998). Parent-of-origin effects on seed development in *Arabidopsis thaliana*. *Development*, 125: 3329–3341.

Segura, S., Scheinvar, L., Olalde, G., Leblanc, O., Filardo, S., Muratalla, A., Gallegos, C. & Flores, C. (2007). Genome sizes and ploidy levels in Mexican cactus pear species *Opuntia* (Tourn.) Mill. series *Streptacanthae* Britton et Rose, *Leucotrichae* DC., *Heliabravoanae* Scheinvar and *Robustae* Britton et Rose. Genet. Resour. *Genetic Resources and Crop Evolution*, 54: 1033–1041.

Simioni, C. & Borges do Valle, C. (2011). Meiotic analysis in induced tetraploids of *Brachiaria decumbens* Stapf. *Crop Breed. App. Biotech.*, 11: 43–49.

Simpson, G.G. (2004). The autonomous pathway: epigenetic and post-transcriptional gene regulation in the control of *Arabidopsis* flowering time. *Current Opinion in Plant Biology*, 7: 570–574.

Soltis, D.E. & Soltis, P.S. (1993). Molecular data and the dynamic nature of polyploidy. *Crit. Rev. Plant Sci.*, 12: 243–273.

Soltis, D.E., Soltis, P.S., Schemske, D.W., Hancock, J.F., Thompson, J.N., Husband, B.C. & Judd, W.S. (2007). Autopolyploidy in angiosperms: have we grossly underestimated the number of species. *Taxon*, 56: 13–30.

Stebbins, G.L. (1950). *Variation and Evolution in plants*. Columbia University Press, New York, USA.

Stebbins, G.L. (1971). *Chromosomal evolution in higher plants*. Columbia University Press, New York

Steimer, A., Schob, H. & Grossniklaus, U. (2004). Epigenetic control of plant development: new layers of complexity. *Current Opinion in Plant Biology*, 7: 11–19.

Strasburger, E. (1900). *Handbook of practical botany*. The Macmillan Co, New York, USA, pp: 294–315.

Tel-Zur, N. (2001) Genetic relationships between vine-cacti of the genera Hylocereus and Selenicereus. In: *Ph.D. Thesis*, Ben-Gurion University of the Negev, Beer-Sheva, Israel.

Tel-Zur, N., Abbo, S., Bar-Zvi, D. & Mizrahi, Y. (2003). Chromosome doubling in vine cacti hybrids. *Journal of Heredity*, 94: 329–333.

Tel-Zur, N., Abbo, S., Bar-Zvi, D. & Mizrahi, Y. (2004). Genetic Relationships among *Hylocereus* and *Selenicereus* Vine Cacti (Cactaceae): Evidence from Hybridization and Cytological Studies. *Annal of Botany*, 94: 527–534.

Tel-Zur, N., Abbo, S. & Mizrahi, Y. (2005). Cytogenetics of semi-fertile triploid and aneuploid intergeneric vine cacti hybrids. *Journal of Heredity*, 96: 124–131.

Tel-Zur, N., Mizrahi, Y., Cisneros, A., Mouyal, J., Schneider, B. & Doyle, J.J. (2011). Phenotypic and genomic characterization of vine cactus collection (Cactaceae). *Genetic Resources and Crop Evolution*, 58: 1075–1085. ISSN 1573-5109.

Tolbert, D.M., Qualset, C.O., Jain, S.K. & Craddock, J.C. (1979). A diversity analysis of a world collection of barley. *Crop Science*, 19:789–794.

Uma, S., Sudha, S., Saraswathi, M.S., Manickavasagam, M., Selvarajam, R., Dukai, P., Sathiamoorthy, S. & Siva, S.A. (2004). Analysis of genetic diversity and phylogenetic relationships among and exotic Silk (AABB) group of bananas using RAPD markers. *J. Hort. Sci. Biotechnol.*, 79: 523–527.

Vigfusson, E. (1970). On polyspermy in the sunflower. *Hereditas*, 64: 1–52.

Von Wangenheim, K.H. (1957). Untersuchungen uber den Zusammenhang swischen. Chromosomenzahl und Kreuzbarkeit bei *Solanum*-Arten. *Z. induckt. Abstamm. u. Vererb. Lehre*, 88: 21–37.

Wakana, S., Iwamasa, M. & Uemoto, S. (1981). Seed development in relation to ploidy of zygotic embryo and endosperm in polyembryonic citrus. *Proc. Int. Soc. Citric*, 1: 35–59.

Wang, C.J., Zhang, L.Q., Dai, S.F., Zheng, Y.L., Zhang, H.G. & Liu, D.C. (2010). Formation of unreduced gametes is impeded by homologous chromosome pairing in tetraploid *Triticum turgidum* × *Aegilops tauschii* hybrids. *Euphytica*, 175: 323–329.

Weiss, J., Nerd, A. & Mizrahi, Y. (1994). Flowering behavior and pollination requirements in climbing cacti with fruit crop potential. *HortScience*, 29: 1487–1492.

Winge, O. (1917). The chromosomes. Their numbers and general significance. *Compt. Rend. Lab. Carlsberg*, 13: 131–275.

Wolfe, K.H. (2001).Yesterday's polyploids and the mystery of diploidization. *Nat. Rev. Genet.*, 2 (5): 333–341.

Yang, R.C., Jana, S. & Clarke, J.M. (1991). Phenotypic diversity and association of some potentially drought-responsive characters in durum wheat. *Crop Science*, 31: 1484–1491.

Zamani, Z., Sarkhosh, A., Fatahi, R. & Ebadi, A. (2007). Genetic relationships among pomegranate genotypes studied by fruit characteristics and RAPD markers. *J. Hort. Sci. Biotechnol.*, 82: 11–18.

Zammouri, J., Arbi, G. & Neffati, M. (2009). Morpho-phenological characterization of *Allium roseum* L. (Alliaceae) from different bioclimatic zones in Tunisia. *African J. Agricul. Res.*, 4 (10): 1004–1014.

Záveský, L., Jarolímová, V. & Štěpánek, J. (2005). Nuclear DNA Content Variation within the Genus *Taraxacum* (Asteraceae). *Folia Geobotanica*, 40: 91–104.

Zhang, L.Q., Liu, D.C., Zheng, Y.L., Yan, Z.H., Dai, S.F., Li, Y.F,, Jiang, Q., Ye, Y.Q. & Yen, Y. (2010). Frequent occurrence of unreduced gametes in *Triticum turgidum–Aegilops tauschii* hybrids. *Euphytica*, 172: 285–294.

Zhang, L.Q., Yen, Y., Zheng, Y.L. & Liu, D.C. (2007). Meiotic restriction in emmer wheat is controlled by one or more nuclear genes that continue to function in derived lines. *Sex. Plant. Reprod.*, 20: 159–166.

8

How Does the Alteration of Meiosis Evolve to Parthenogenesis? - Case Study in a Water Flea, *Daphnia pulex*

Chizue Hiruta[1] and Shin Tochinai[1,2]
*[1]Department of Natural History Sciences, Graduate School of Science,
Hokkaido University,
[2]Department of Natural History Sciences, Faculty of Science, Hokkaido University,
Japan*

1. Introduction

This chapter reviews progress in our understanding of two different reproductive modes in *Daphnia pulex* and discusses the interesting modifications found in meiosis during parthenogenesis.

Most daphnid species reproduce parthenogenetically as well as sexually, resulting in the production of diploid progenies in both cases. In natural populations, parthenogenesis is the common mode of reproduction, and parthenogenetic offspring are all female. However, in response to certain environmental conditions, such as crowding or seasonal change, male offspring are also produced parthenogenetically, and sexual reproduction occurs (Hebert, 1978). Although they switch between parthenogenesis and sexual reproduction in response to environmental conditions, little is known about the molecular and cytological mechanisms switching and governing each reproductive mode. It can be interpreted that *D. pulex* develops a reproductive strategy utilizing 'parthenogenesis', which has high reproductive power, and 'sexual reproduction', which generates genetic diversity, in response to different environments. These theoretical studies have been made on evolutionary mechanism of reproductive modes (Decaestecker *et al.*, 2009); however, practically no study analyzing the evolutionary mechanism while taking the developmental constraints into consideration has so far been conducted. Our understanding of the evolution of reproductive strategy would increase once we precisely clarify the developmental gene programs operating there. We have recently started to develop *D. pulex* as an experimental model for studying oogenesis and developmental mechanisms during evolution.

1.1 Chapter contents

- At first, we explain why we are interested in the comparative research of two reproductive modes: parthenogenesis and sexual reproduction, and why we chose *D. pulex* as an experimental animal.

- From our experiments, it was concluded that diploid progeny are produced by non-reductional division in parthenogenetic *D. pulex*. We found that, when a parthenogenetic egg entered the first meiosis, division was arrested in the early first anaphase. Then, two half-bivalents, which were dismembered from each bivalent, moved back to the equatorial plate and assembled to form a diploid equatorial plate. Finally, the sister chromatids were separated and moved to opposite poles in the same manner as the second meiotic division.

- We hypothesized that *D. pulex* switches reproductive mode by controlling the maturation division of oocytes (arrest or progress), depending on whether the egg is fertilized or not.

- We use *D. pulex* as an example of the evolutionary process of parthenogenesis which arose from sexual reproduction. Suggestions for future research are also presented.

2. A water flea, *Daphnia pulex*

Water fleas of the genus *Daphnia* are members of the order Cladocera, which are small crustaceans that live in various aquatic environments varying from temporary ponds to large lakes as a cosmopolitan species. They occupy a key position (food chain) in aquatic communities, both as important herbivores eating algae and bacteria and as major prey items of fish, aquatic insects and other predators. Moreover, they have a value as environmental indicator organisms because of their high sensitivity to water quality. Thus, many previous studies have concentrated on ecology, taxonomy and toxicology. Recently however, the spotlight has been on the phenotypic plasticity of *D. pulex*. One example of this phenomenon is that they express morphological, life history, and behavioral defenses in response to chemical cues released by predators (Tollrian & Dodson, 1999).

2.1 General feature

2.1.1 Anatomical characteristics

The carapace is transparent and encloses the whole trunk, except the head and the apical spine. The trunk appendages are flattened, leaf-like structures that serve for suspension feeding and for locomotion. There is a single central compound eye in the head. The large, paired appendages used for swimming are second antennae.

Female: The body length of the female is about 1-3 mm. The first antennae are small and short, not extending beyond the rostrum. The ovaries are a pair of elongated organs lying on either side of the gut. In the posterior part of the each ovary, the oviduct opens into the brood chamber. As shown in Fig. 1C, oogonia and smaller oocytes are located in the most-posterior part of the ovary, and move anteriorly as development progresses. The number of eggs spawned in a clutch depends on the nutritional state and the size of the female. The space between the body and the carapace is used as a brood chamber.

Male: The males are smaller than the females. The rostrum is generally indistinct and the first antennae are large and long. The male has a copulatory hook, which is used for holding on to the female during mating, on the first thoracic leg. The testes are a pair of elongated organs lying on either side of the gut. The two gonopores open near the anus. As shown in Fig. 1F, the mature testis has spermatozoa in the lumen. Spermatogenesis begins at the walls and proceeds into the innermost part (lumen).

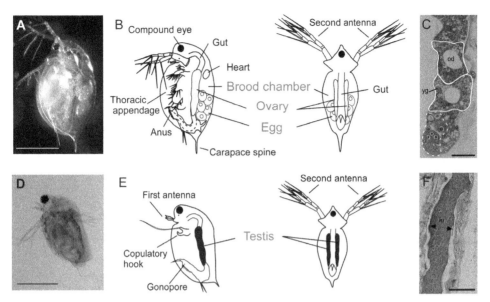

Fig. 1. *Daphnia pulex*. Anatomical characteristics of an adult female (top row) and male (bottom row). (A,D) Photo. Scale bars = 1 mm in A; 0.5 mm in D. (B,E) Lateral (left) and frontal (right) view of anatomy. (C) Sagittal section of the ovary. The largest eggs (solid white line) contain a large amount of yolk granules and oil droplets. Oogonia and smaller oocytes are located in the most-posterior part of the ovary (dashed white line). Scale bar = 50 μm. yg, yolk granule; od, oil droplet. (F) Sagittal section of the testis. Sperms are located in the lumen. Scale bar = 100 μm. lu, lumen.

2.1.2 Life cycle

Daphnia pulex reproduce either by parthenogenesis or sexual reproduction and populations are almost exclusively female. Under favorable conditions, eggs are produced in clutches of one to several dozen, and one female may produce several clutches, which is linked with the molting process. The eggs are laid in the brood chamber shortly after molting. Embryonic development occurs in the brood chamber and the larvae are miniature versions of the adults. The neonates are released from the brood chamber just before the mother molts. In this way, the parthenogenetic individual repeats the cycle of molting, egg laying and releasing during her life (Fig. 2, Non-resting cycle).

Unfavorable conditions, such as changes in water temperature or food deprivation as a result of population increase, may induce the production of males. In other words, a single parthenogenetic female can produce either parthenogenetic female offspring and/or males for sexual reproduction. The male clasps the female and display copulatory behavior. Then the resting eggs are produced (Fig. 2, Resting cycle). Only two large eggs produced in a single clutch (one from each ovary) are enclosed in an ephippium which used to be a part of the dorsal exoskeleton and is darkly pigmented with melanin. The resting eggs are resistant to desiccation and freezing during winter, playing an important role in colonizing new habitats or in the re-establishment of an extinguished population after unfavorable conditions.

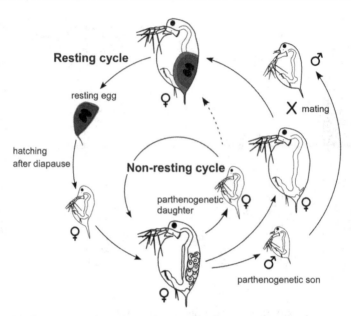

Fig. 2. Life cycle of *D. pulex*. Life cycle can be divided into 'resting cycle' and 'non-resting cycle'. In natural populations, parthenogenesis is the common mode of reproduction, and parthenogenetic offspring are normally female (parthenogenetic non-resting cycle). However, in response to unfavorable conditions, such as crowding or seasonal change, male offspring are also produced parthenogenetically, and then sexual reproduction occurs (sexual resting cycle). Although a female usually produces resting eggs requiring fertilization by sperm, it sometimes happens that a female produces parthenogenetic resting eggs (parthenogenetic resting cycle, dashed arrow).

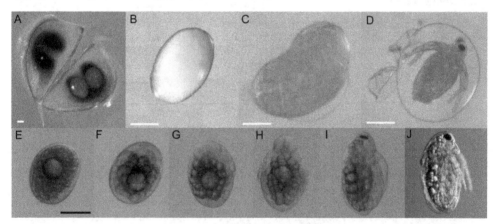

Fig. 3. Embryogenesis of resting egg (top row) and non-resting egg (bottom row). (A-D) The resting egg was produced by parthenogenesis. There was no difference in the manner of development between a parthenogenetic resting egg and a sexual resting egg. Scale bars = 100 μm. (E-J) Embryonic development was completed in about 3 days at 18°C. Scale bar = 100 μm.

There are two types of eggs, i) **resting egg**, also termed 'winter egg', 'diapause egg', 'dormant egg' or 'ephippial egg', ii) **non-resting egg**, also known as 'summer egg' or 'subitaneous egg', as mentioned above (Figs. 2 and 3). In any maturation stage they are easily distinguishable from each other in the ovary of the living animal. Although the non-resting eggs produce both female and male neonates, the resting eggs produce female offspring without exception.

Examination of the literature on reproduction of this species reveals a great deal of confusion regarding the relationship between the types of egg and reproductive modes. Interestingly, Hobæk & Larsson (1990) reported that the formation of male offspring and resting eggs are independently controlled, possibly by distinct sets of environmental cues. We have also observed parthenogenetic resting eggs in *D. pulex* (Fig. 2, dashed arrow) and confirmed that the eggs have the potential to develop normally (Fig. 3A-D). Most daphnids are believed to switch to sexual reproduction resulting in the production of resting eggs; however, several daphnid species, not forgetting *D. pulex*, can produce them without fertilization. Moreover, the switch to sexual reproduction involves a commitment by the mother to produce a resting egg case (ephippium) and to produce eggs capable of diapause. Thus, the decision to produce a non-resting or a resting egg must be made long before the point at which the egg is fertilized. So far, it is clear that 1) both non-resting and resting eggs are produced by parthenogenesis, 2) resting eggs are also produced by sexual reproduction. Whether non-resting eggs can be produced by sexual reproduction or not has never been studied.

2.2 *Daphnia pulex* as an evolutionary developmental biology (evo-devo) model

Daphnia pulex is an ideally suited laboratory animal for workers in the fields of development and genetic research: It is easy to raise under laboratory conditions, propagates quickly because of its short reproductive cycle, and can induce male offspring by a juvenile hormone, methyl farnesoate (Olmstead & Leblanc, 2002). In addition, new experimental techniques (e.g., *in situ* hybridization, immunofluorescence, microinjection, RNAi) were established in daphnid species over the past several years (Sagawa *et al.*, 2005; Tsuchiya *et al.*, 2009; Kato *et al.*, 2011). Moreover, a recent description of the complete genome sequence for *Daphnia pulex* (Colbourne *et al.*, 2011) and genetic linkage map (Cristescu *et al.*, 2006) will provide us with a powerful tool for analyzing the molecular mechanism of any aspect in this species, including enigmatic meiotic processes. *Daphnia pulex* has only a 200-megabase genome and as many as about 31,000 genes. Thirty-six percent of *Daphnia pulex* genes have no detectable homologs with other animal species and about 13,000 genes have been identified as paralogs. There will be a good chance to find novel genes which enable the evolution of a unique reproductive strategy in this species. For these reasons, *D. pulex* started to garner attention as an evo-devo model animal (Jenner & Wills, 2007).

3. Reproductive mode of *D. pulex*

Daphnia pulex adopt parthenogenesis and sexual reproduction differentially in response to varied environmental cues as mentioned above. The production of diploid progenies is a common finding in both reproductive modes.

3.1 Parthenogenesis

The reproductive modes of *D. pulex* were studied over the past century. Previous studies suggested that *D. pulex* produces parthenogenetic eggs via apomixis; the nuclear division of mature oocytes should be an equational division equal to somatic mitosis (Kühn, 1908; Ojima, 1954, 1958; Zaffagnini & Sabelli, 1972). However, due to the presence of a large amount of yolk and the minute size of chromosomes, it was not easy to observe the nucleus during the process of oogenesis and therefore the behavior of chromosomes in the ovarian egg remained undescribed. In spite of previous reports that suggested the occurrence of mitosis in parthenogenetic oocytes, we found "abortive meiosis" instead during the oogenesis of parthenogenetic *D. pulex*, as mentioned below (see section 5). This finding suggests that parthenogenetic *D. pulex* may switch to sexual reproduction by progressing the maturation division of oocytes from arrest. Moreover, it would give parthenogenetic *D. pulex* a mechanism making recombination possible even under parthenogenesis. In other words, it can lead to offspring with genetic variability because chromosomal recombination can take place between homologous chromosomes, while there is no introduction of new genes from another individual.

3.2 Sexual reproduction

At present very little is known about the process of meiosis and fertilization during sexual reproduction in this species. Although it still is a matter of debate, Ojima (1958) reported that sperm seemed to penetrate into the ovarian egg. The structure of the spermatozoa is very atypical in daphnid species. The mature sperm of *D. pulex* lacks a flagellum and is therefore not actively mobile. Indeed, during mating, the male deposits sperm near the openings of the female gonopore. Studies are needed to clarify the timing of fertilization and the process of meiosis. It will provide us not only with insight into the switching mechanism of reproductive modes, but also with the differences between normal meiosis and abortive meiosis.

4. Oogenesis of parthenogenetic *D. pulex*

In this section, we offer a close overview of the growth and maturation of the parthenogenetic eggs. Mature and spawned eggs in the same brood were the same size and mostly synchronized in the maturation stage (cell cycle). It takes approximately 60 h for oocytes to grow in the ovary (from 0 to 60 h, Fig. 4). Inclusions such as yolk granules and oil droplets increased in size and number until the fully-grown egg is formed. The nuclear division apparatus appears just after molting (at 0 min after molting (0 AM): Fig. 4). We observed the precise states of nuclei in the eggs with the following timing during the course of parthenogenesis (from egg maturation to early development): 1) the time of molting of the female (Fig. 4, 0 AM), 2) the interval between molting and 13 min after molting. The parthenogenetic eggs began to migrate from the ovary to the brood chamber and this process was completed within about 3 min, 3) the time when oviposition was completed (Fig. 4, 0 min post oviposition (0 PO)), and 4) the time during which the parthenogenetic eggs in the brood chamber began to develop.

At the stage of 0 h of egg maturation, the oocyte was at first morphologically indistinguishable from the nurse cell (Fig. 5A). In early stages, the development of both oocytes and nurse cells proceeds in a similar manner, but only the egg cells developed to form yolk granules and oil droplets during maturation, whereas the nurse cells became smaller in size and finally degenerated (Fig. 5B-D). With the passage of time, the yolk granules and oil droplets increased in number in the oocytes. Soon after grown juveniles were discharged from the brood chamber, the nuclear membrane of the fully-matured ovarian egg to be spawned gradually disappeared, and finally the breakdown of the germinal vesicle took place just before molting of the mother (Fig. 5E).

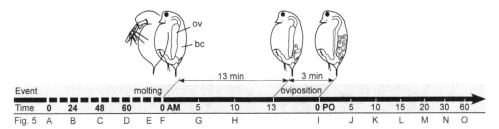

Fig. 4. The time course of parthenogenesis in *D. pulex*. The oviposition cycle is about 3 days at 18°C. The broken line of the time axis (from 0 to 60) indicates the time scale in hours, and the solid line of time axis (from 0 AM to 60 PO) indicates the time scale in minutes. Usually, after releasing all neonates that developed in the dorsal brood chamber, females molt. After this molt occurs at 0 AM (minutes after molting), a strict time course proceeds. The female begins extruding eggs into the brood chamber at 13 AM and this process is completed within about 3 min. The point of 0 PO (minutes post oviposition) indicates the time when the female extruded the last egg. Then, the parthenogenetic eggs in the brood chamber develop into juveniles. Ovarian and spawned eggs in a clutch are approximately the same size and mostly synchronized in the cell cycle. In the lower part of Fig. 4, the capital letters show the point of time when the specimens were observed in Fig. 5. ov, ovary; bc, brood chamber. Modified from Hiruta *et al.*, 2010.

The chromosomes co-oriented in a position midway between the poles, and then each bivalent started to separate into two half-bivalents, one moving to each pole of the spindle by 5 AM (Fig. 5F, G). However, the movement of chromosomes from the metaphase plate to the poles was arrested at an early stage of anaphase before 10 AM (Fig. 5H). Egg laying (oviposition) began at 13 AM. The migration of all eggs from the ovary to the brood chamber was completed within about 3 min (= 0 PO). At 0 PO, the chromosomes moved back and assembled as a diploid equatorial plate around the equator of the spindle in the spawned egg (Fig. 5I). By 5 PO, the division apparatus migrated to the periphery and the cell division cycle restarted. Then the sister chromatids moved apart, one going to each pole of the spindle through metaphase and anaphase by 10 PO (Fig. 5J, K). The complete set of chromosomes was lifted above the egg surface and eventually one polar body-like small daughter cell was extruded at around 20 PO (Fig. 5L, M).

After the completion of oogenesis, the chromosomes left in the egg moved deeper inside the egg (Fig. 5N) and mitosis occurred without cytokinesis, resulting in a polynuclear syncytial embryo (Fig. 5O). Then, the nuclei migrated to the periphery, and a typical superficial cleavage proceeded.

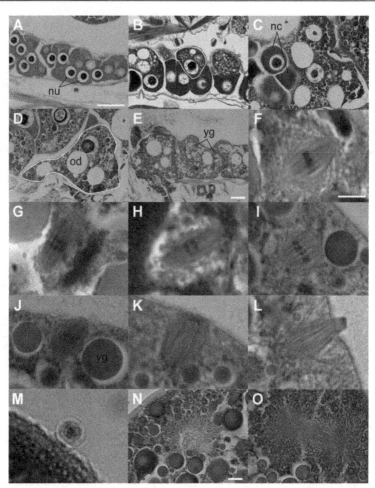

Fig. 5. Oogenesis of parthenogenetic *D. pulex*. (A-E) Oocytes and nurse cells in the ovary. (F-H) The division apparatus in ovarian eggs. (I-O) The division apparatus in eggs spawned to the brood chamber. (A) 0 h. The oocytes and nurse cells were indistinguishable. (B) 24 h. Yolk and oil droplet formation took place only in oocytes. (C) 48 h. The yolk granules and oil droplets increased. (D) 60 h. The degeneration of nurse cells proceeded. (E) Just before molting. The breakdown of the germinal vesicle (GVBD) took place. (F) 0 AM. Division apparatus appeared and chromosomes aligned at the metaphase plate. (G) 5 AM. Each bivalent is separated into two half-bivalents. (H) 10 AM. The division seemed to stop at early anaphase. (I) 0 PO. The chromosomes moved back and rearranged around the equator of the spindle. (J) 5 PO. The chromosomes started to separate. (K) 10 PO. The division proceeded to anaphase. (L) 15 PO. One complete set of chromosomes was lifted above the egg surface. (M) 20 PO. A polar body-like small daughter cell was extruded. (N) 30 PO. The swelled chromosomes left in the egg moved to a deeper part. (O) 60 PO. The first cleavage proceeded without cytokinesis. Scale bars = 100 μm in A-E; 5 μm in F-M; 10 μm in N and O. nc, nurse cell; nu, nucleus; od, oil droplet; yg, yolk granule. Solid white circle indicates an oocyte. Modified from Hiruta *et al.*, 2010 and unpublished data.

5. Abortive meiosis found in parthenogenetic *D. pulex*

Parthenogenetic eggs achieve two successive divisions like normal meiosis while the first division is abortive (see section 4). In the first meiosis, bivalents align at the equatorial plate (Fig. 6A) and begin to separate into two half-bivalents (Fig. 6B). Then, each half-bivalent moves back and sister chromatids rearrange as a diploid equatorial plate around the equator of the spindle (Fig. 6C). Finally, the second meiosis-like division takes place normally, producing a single polar body-like extremely small daughter cell (Fig. 6D). Compared with meiosis, it is known that the first meiotic division is skipped there.

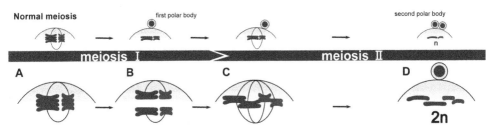

Fig. 6. Schematic illustration of parthenogenesis in *D. pulex*. The top row shows the process of meiosis. In the first meiosis, bivalents aligned at the metaphase plate and separate into two half-bivalents, producing the first polar body. Then, the second meiosis takes place, producing the second polar body. As a result, a haploid egg is produced. By contrast, a diploid egg is produced by parthenogenesis in *D. pulex*. (A) In the first meiosis, bivalents align at the equatorial plate and begin to separate into two half-bivalents. (B) However, the division is arrested at early anaphase. (C) Then each half-bivalent moves back and sister chromatids rearrange as a diploid equatorial plate around the equator of the spindle. (D) Finally, the second meiosis-like division takes place normally, producing a single polar body-like extremely small daughter cell. Illustration adopted from Hiruta *et al.*, 2010.

5.1 Hypothetical model for reproductive strategy in *D. pulex*

We hypothesized that *D. pulex* switches its reproductive mode (sexual or parthenogenetic) depending on whether the egg is fertilized or not. It is highly plausible that, if the egg is not fertilized, the first meiosis is aborted and, subsequently, a second meiosis-like division takes place as observed in our study. On the other hand, normal meiosis may well occur if the ovarian egg is fertilized. If this is true, it seems appropriate to assume that fertilization occurs at the stage between first metaphase and anaphase in the ovarian egg.

Since the two reproductive modes (sexual reproduction or parthenognesis) are not strictly associated with the two types of egg (resting or non-resting egg) as mentioned in section 2.1.2, a study will need to be conducted to verify the hypothesis including whether the non-resting egg is produced by sexual reproduction. If this hypothesis is true, daphnid species adopt the least waste system in the production of eggs, because eggs are able to develop regardless of fertilization. In order to verify this hypothesis, we are currently trying to establish the methods for *in vitro* maturation of ovarian eggs and artificial insemination.

6. Spindle assembly and spatial distribution of γ-tubulin during abortive meiosis in parthenogenetic *D. pulex*

During the analysis of nuclear maturation division, which resulted in either parthenogenetic or sexual development of the egg, we found that the spindle in abortive meiosis was barrel-shaped, anastral, and organized without centrosomes (Fig. 7, Hiruta C., unpublished). Corresponding to our results, a barrel-shaped meiotic spindle without centrosomes has been reported in several animal species including parthenogenetic pea aphid (Riparbelli *et al.,* 2005). The centrosomes are present in cleavage division (mitosis) of both reproductive modes. In the case of sexual reproduction in various animals, the sperm supplies the centriole after fertilization. On the other hand, in the case of parthenogenesis in insects, centrosomes are spontaneously assembled in mitotic spindle microtubules. We suspect that *de novo* assembled centrosomes could be an evolutionary conserved process leading to parthenogenetic development.

Even more surprisingly, gamma (γ)-tubulin is localized along spindle microtubules for the duration of abortive meiosis, while it is present only on the centrosomes in parthenogens' cleavage division (Fig. 7, Hiruta C., unpublished). The results from *D. pulex* are identical with those from pig oocytes, but not universal among animals (Lee *et al.,* 2000). Incidentally, the localization of γ-tubulin to centrosomes corresponds to a typical spindle formation which is highly conserved in animals. Comparative research needs to be conducted to reveal whether sperm entry affects the localization of γ-tubulin in *D. pulex* or not.

(A) Abortive meiosis   (B) Cleavage division (mitosis)
α -tubulin   γ -tubulin   α -tubulin   γ -tubulin

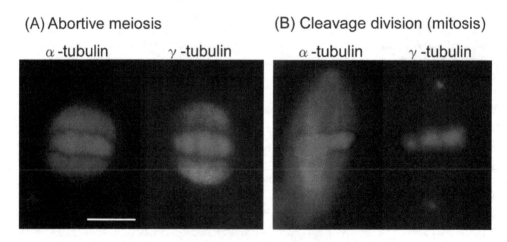

Fig. 7. Immunofluorescence localization of α- and γ-tubulin during abortive meiosis and cleavage division (mitosis). Scale bar = 5 μm. Chromosomes were counter-stained with DAPI (blue). (A) Spindle formation was barrel-shaped, anastral, and organized without centrosomes. Gamma-tubulin was distributed along spindle microtubules during abortive meiosis. (B) Spindle formation was spindle-shaped, astral, and had organized centrosomes. Gamma-tubulin was present only the spindle poles. Hiruta C., unpublished.

7. Molecular basis of the transition from sexual reproduction to parthenogenesis

There are variations of the meiotic program that would produce diploid oocytes by skipping a division, by fusion of a haploid oocyte with a polar body, or by premeiotic endoreplication of chromosomes (Suomalainen et al., 1987; Schön et al., 2009). At any rate, it is thought that parthenogenesis evolved from sexual reproduction by changing the meiotic program. It seems likely that a relatively simple deviation from the established program of oogenesis, i.e., meiosis, is sufficient to permit parthenogenesis. Schurko et al. (2009) reported that expression patterns of meiosis-related and meiosis-specific genes (e.g., SMCs, REC8) during sexual reproduction of D. pulex are similar to that during parthenogenesis. Some of these genes are present in multiple copies and might be expressed differently in sexual reproduction and parthenogenesis. To cite one example, there are seven RECQ2 copies, which limit crossing over, which could have evolved novel roles in parthenogenesis. In fact, as if to correspond to the expectation, we have so far failed to observe chiasmata where crossing over occurred. In addition, it is suggested that PLK1, which is involved in orienting kinetochores during mitosis and meiosis, controls the localization within the spindle of γ-tubulin for mitotic spindle formation through the augmin complex in human and drosophila cells (Goshima et al., 2008; Uehara et al., 2009). As mentioned in section 6, the γ-tubulin is localized along the spindle in abortive meiosis. There is a possibility that a PLK copy is expressed in parthenogenesis specifically to operate the localization of γ-tubulin. We expect to reveal how the meiotic program is altered resulting in parthenogenesis. On the other hand, the absence of meiosis-specific DMC1 suggests that innovations for recombination in meiosis and parthenogenesis in D. pulex may have evolved. In the last paragraph, we stated factors associated with regulation of chromosomal organization. The following discussion is about the cell cycle control factors. There are several copies of cell cycle proteins, such as cyclins, cdks and polo kinases in D. pulex. In particular, the Mos-MAPK (mitogen-activated protein kinase) pathway, which is responsible for metaphase arrest in meiosis I or II before fertilization in many animal species (Sagata, 1996), might be a candidate for cell cycle arrest in parthenogenetic D. pulex. We have preliminary data that MAP kinase is expressed in D. pulex oocyte in metaphase of meiosis I.

Consequently, a precise description of the reproductive modes in D. pulex allows us to understand the mechanism which is widely preserved in eukaryotes and in the Daphnia-specific mechanism, namely, commonality and diversity of the pattern of division.

8. How does the transition from sexual reproduction to parthenogenesis occur during evolution?

Parthenogenesis, which is thought to arise from the alteration of meiosis, has been found in various animal species (Suomalainen et al., 1987; Schön et al., 2009). In many taxonomic groups, parthenogenesis was independently acquired during evolution. By comparing those cases, we could gain insight into the transition from sexual reproduction to parthenogenesis, many of which resulted from the change made during meiosis. According to Suomalainen et al. (1987), the following cases have been categorized as the type of 'skipped the first meiosis' (Fig. 8).

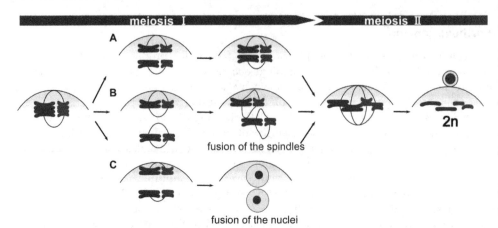

Fig. 8. Different types of abortive meiosis. These are categorized as 'skipped the first meiosis'. (A) *Daphnia pulex*. Two half-bivalents move back to the equatorial plate. (B) *Apterona helix*. Two spindles, an inner and outer, come together side by side and fuse. (C) *Fasciola hepatica*. Two haploid nuclei fuse.

In Lepidoptera *Apterona helix* (Fig. 8B), the first meiosis is aborted at the end of anaphase and two metaphase plates, an inner and outer, are formed. Both of these contain the haploid number of chromosomes and have separate spindles. Before the second meiosis, the inner metaphase spindle with its chromosome plate moves to the side of the outer spindle. Finally, the two spindles and the metaphase plates lie side by side and fuse (Narbel, 1946). In Crustacea *Artemia salina*, Lepidoptera *Solenobia lichenella*, *Luffia ferchaultella* and *L. lapidella*, meiosis is interrupted at some stage between the end of the first anaphase and the second metaphase (Narbel-Hofstetter, 1950, 1963, 1965; Stefani, 1960). Then the two haploid plates reunite, forming a new metaphase spindle, and the second diploid meiosis is accomplished. In Trematoda *Fasciola hepatica* (Fig. 8C), the first meiosis occurs without cytokinesis, giving rise to two haploid nuclei. These nuclei fuse and form a diploid cleavage nucleus (Sanderson, 1952).

There are differences at the stage of division arrest and restoration or maintenance of diploidy even in the same division that skipped the first meiosis. At any rate, if first meiosis is completely skipped, parthenogenetic division becomes congruent with mitosis and finally turns out to be obligate parthenogenesis. A comparative and detailed study of the parthenogenetic mechanism will bring us to a better understanding of the possible alteration of meiosis during evolution.

9. Conclusion

In conclusion, *D. pulex* is suitable for experimentation to understand the evolution of reproductive modes from a viewpoint of evolutionary developmental biology. The case study in *D. pulex* will contribute to a discussion of challenges such as how does parthenogenesis work and how evolution of sexual reproduction and parthenogenesis occurs.

10. Acknowledgments

We are grateful to the members of the Tochinai laboratory for their helpful advice and discussion. We also thank a great number of *D. pulex* that made a sacrifice for our research. CH was supported by a JSPS Research Fellowships for Young Scientists (No. 09J01653). This work was supported by a Grant-in-Aid for Exploratory Research, financed by the Ministry of Education, Culture, Sports, Science and Technology, Japan to ST (No. 21657054).

11. References

Colbourne, JK., Pfrender, ME., Gilbert, D., Thomas, WK., Tucker, A., Oakley, TH., Tokishita, S., Aerts, A., Arnold, GJ., Basu, MK., Bauer, DJ., Cáceres, CE., Carmel, L., Casola, C., Choi, JH., Detter, JC., Dong, Q., Dusheyko, S., Eads, BD., Fröhlich, T., Geiler-Samerotte, KA., Gerlach, D., Hatcher, P., Jogdeo, S., Krijgsveld, J., Kriventseva, EV., Kültz, D., Laforsch, C., Lindquist, E., Lopez, J., Manak, JR., Muller, J., Pangilinan, J., Patwardhan, RP., Pitluck, S., Pritham, EJ., Rechtsteiner, A., Rho, M., Rogozin, IB., Sakarya, O., Salamov, A., Schaack, S., Shapiro, H., Shiga, Y., Skalitzky, C., Smith, Z., Souvorov, A., Sung, W., Tang, Z., Tsuchiya, D., Tu, H., Vos, H., Wang, M., Wolf, YI., Yamagata, H., Yamada, T., Ye, Y., Shaw, JR., Andrews, J., Crease, TJ., Tang, H., Lucas, SM., Robertson, HM., Bork, P., Koonin, EV., Zdobnov, EM., Grigoriev, IV., Lynch, M. & Boore, JL. (2011). The ecoresponsive genome of *Daphnia pulex*. *Science*, Vol.331, No.6017, pp.555-561

Cristescu, MEA., Colbourne, JK., Radivojac, J. & Lynch, M. (2006). A microsatellite-based genetic linkage map of the waterflea, *Daphnia pulex*: On the prospect of crustacean genomics. *Genomics*, Vol.88, No.4, pp.415-430

Decaestecker, E., Meester, LD. & Mergeay, J. (2009). Cyclical parthenogenesis in *Daphnia*: Sexual versus asexual reproduction, In: *Lost sex: The evolutionary biology of parthenogenesis*, Schön, I., Martens, K. & Dijk, PV., pp.295-316, Springer, ISBN 978-90-481-2769-6

Goshima, G., Mayer, M., Zhang, N., Stuurman, N. & Vale, RD. (2008). Augmin: a protein complex required for centrosome-independent microtubule generation within the spindle. *J Cell Biol.*, Vol.181, No.3, pp.421-429

Hebert, PDN. (1978). The population biology of *Daphnia* (Cruatacea, Daphnidae). *Biol. Rev.*, Vol.53, pp.387-426

Hiruta, C., Nishida, C. & Tochinai, S. (2010). Abortive meiosis in the oogenesis of parthenogenetic *Daphnia pulex*. *Chromosome Res.*, Vol.18, No.7, pp.833-840

Hobæk, A. & Larsson, P. (1990). Sex determination in *Daphnia magna*. *Ecology*, Vol.71, No.6, pp.2255-2268

Jenner, RA. & Wills, MA. (2007). The choice of model organisms in evo-devo. *Nat Rev Genet*, Vol.8, No.4, pp.311-319

Kato, Y., Shiga, Y., Kobayashi, K., Tokishita, S., Yamagata, H., Iguchi, T. & Watanabe, H. (2011). Development of an RNA interference method in the cladoceran crustacean *Daphnia magna*. *Dev Genes Evol*, Vol.220, pp.337-345

Kühn, A. (1908). Die Entwicklung der Keimzellen in den parthenogenetischen Generationen der Cladoceren *Daphnia pulex* de Geer und *Polyphemus pediculus* de Geer. *Arch Zellforsch*, Vol.1, pp.538-586

Lee, J., Miyano, T. & Moor, RM. (2000). Spindle formation and dynamics of γ-tubulin and nuclear mitotic apparatus protein distribution during meiosis in pig and mouse oocytes. *Biol Reprod*, Vol.62, No.5, pp.1184-1192

Narbel, M. (1946). La cytologie de la parthénogenèse chez *Apterona helix* Sieb. (Lepid, Psychides). *Rev. Suisse Zool*, Vol.53, pp.625–681

Narbel-Hofstetter, M. (1950). La cytologie de la parthénogenèse chez *Solinobia* sp. (*lichenella* L.?) (Lépidoptères, Psychides). *Chromosoma*, Vol.4, pp.56–90

Narbel-Hofstetter, M. (1963). Cytologie de la pseudogamie chez *Luffia lapidella* Goeze (Lepidoptera, Psychidae). *Chromosoma*, Vol.13, pp.623–645

Narbel-Hofstetter, M. (1965). La variabilité cytologique dans la descendance des femelles de *Luffia ferchaultella* Steph. (Lepidoptera, Psychidae). *Chromosoma*, Vol.16, pp.345–350.

Ojima, Y. (1954). Some cytological observations on parthenogenesis in *Daphnia pulex* (de Geer). *Jour Fac Sci Hokkaido Univ, Ser.VI, Zool.12*, pp.230–241

Ojima, Y. (1958). A cytological study on the development and maturation of the parthenogenetic and sexual eggs of *Daphnia pulex* (Crustacea, Cladocera). *Kwansei Gakuin Univ*, Annual Studies 6, pp.123–176

Olmstead, AW. & Leblanc, GA. (2002). Juvenoid hormone methyl farnesoate is a sex determinant in the crustacean *Daphnia magna*. *J Exp Zool*, Vol.293, No.7, pp.736–739

Riparbelli, MG., Tagu, D., Bonhomme, J. & Callaini, G. (2005). Aster self-organization at meiosis: a conserved mechanism in insect parthenogenesis?. *Dev Biol*, Vol.278, No.1, pp.220–230

Sagata, N. (1996). Meiotic metaphase arrest in animal oocytes: its mechanisms and biological significance. *Trends Cell Biol*, Vol.6, No.1, pp.22–28

Sagawa, K., Yamagata, H. & Shiga, Y. (2005). Exploring embryonic germ line development in the water flea, *Daphnia magna*, by zinc-finger-containing VASA as a marker. *Gene Expr Patterns*, Vol.5, pp.669–678

Sanderson, AR. (1952). Maturation and probable gynogenesis in the liver fluke, *Fasciola hepatica* L.. *Nature*, Vol.172, pp.110–112

Schön, I., Martens, K. & Dijk, PV. (2009). *Lost sex: The evolutionary biology of parthenogenesis*. Springer, ISBN 978-90-481-2769-6

Schurko, AM., Logsdon, JM Jr. & Eads, BD. (2009). Meiosis genes in *Daphnia pulex* and the role of parthenogenesis in genome evolution. *BMC Evol Biol*, Vol.9, No.78

Stefani, R. (1960). L'*Artemia salina* parthenogenetica a Cagliari. *Riv Biol*, Vol.52, pp.463–490

Suomalainen, E., Saura, A. & Lokki, J. (1987). *Cytology and evolution in parthenogenesis*. CRC Press, ISBN 0-8493-5981-3, Boca Raton, FL

Tollrian, R. & Dodson, SI. (1999). Inducible defenses in Cladocera: Constraints, costs, and multipredator environments, In: *The ecology and evolution of inducible defenses*, Tollrian, R. & Harvell, CD., pp.177–202, Princeton, ISBN 0-691-00494-3, UK

Tsuchiya, D., Eads, BD. & Zolan, ME. (2009). Methods for meiotic chromosome preparation, immunofluorescence, and fluorescence *in situ* hybridization in *Daphnia pulex*, In: *Meiosis, Volume 2: Cytological Methods, vol. 558*, Keeney, S., pp.235–249, Springer, ISBN 978-1-60761-103-5, USA

Uehara, R., Nozawa, RS., Tomioka, A., Petry, S., Vale, RD., Obuse, C. & Goshima, G. (2009). The augmin complex plays a critical role in spindle microtubule generation for mitotic progression and cytokinesis in human cells. *Proc Natl Acad Sci*, Vol.106, No.17, pp.6998–7003

Zaffagnini, F. & Sabelli, B. (1972). Karyologic observations on the maturation of the summer and winter eggs of *Daphnia pulex* and *Daphnia middendorffiana*. *Chromosoma*, Vol.36, No.2, pp.193–203

9

Avian Meiotic Chromosomes as Model Objects in Cytogenetics

Katarzyna Andraszek and Elżbieta Smalec
Siedlce University of Natural Sciences and Humanities,
Poland

1. Introduction

The beginning of the new millennium proved critical for life sciences. The team of scientists working on the Human Genome Mapping Project published and made available on the website of *Nature* a complete set of information on the human genome. Almost simultaneously an independent private company, Celera Genomics, published the results of its study of human DNA sequence in *Science*. The dream of scientists trying to read the human genome sequence for the last ten years was fulfilled. The end of the related rivalry was the beginning of a new epoch and a new field in genetics – genomics.

At present, genomics and cytogenetics are the two fastest-developing genetic disciplines. Genome mapping and karyotype standardization projects involve an increasingly greater number of animal and plant species. Genome exploration by physical mapping requires a knowledge of the karyotype of a given species for which the map is being created. The gene/chromosome interdependence is investigated through cooperation between international mapping projects and international karyotype standardization programs. A particularly dynamic rate of progress is observed in mapping the genomes of higher vertebrates. The most numerous group of vertebrates is constituted by birds and, as a paradox, it is their genome that is least well-known. Although they guard the secrets of their karyotype with the high diploid number and microchromosomes, birds have meiotic chromosomes of easy access and the intriguing lampbrush chromosomes.

The high number of chromosomes and the presence of microchromosomes in the avian karyotype have made cytogeneticists look for other sources of information on chromosomes. The extension of cytogenetic investigations over a greater number of chromosomes required the use of chemical agents, such as amethopterin or thymidine that inhibit chromosome condensation. Such experiments were undertaken on hens and produced poorly condensed chromosomes with bands in fine resolution. This made it possible to identify sixteen chromosome pairs.

Meiotic chromosomes are a valuable object for avian karyotype analyses. Meiosis is normally observed in males, as spermatocytes are relatively small, numerous and readily available. The most often analysed meiotic chromosomes are those contained in cells in the pachytene and diplotene of the first meiotic division. The experiments have predominantly concerned samples of testes. The observation of female meiotic chromosomes is limited by the large size of the egg cell, technical difficulty in sampling chromosomes from the egg cell and low numbers of the cells as compared with spermatocytes. Nevertheless, female meiotic chromosomes are worth paying attention to. The maturing oocytes isolated form the ovaries constitute the material for this type of experiments. Meiosis guarantees the stability of the chromosome number in the consecutive generations of sexually reproducing organisms. Generally, meiosis is represented as a process during which the cell passes through totally different stages that correspond with particular structure and behaviour of chromosomes. The observation of the changes occurring in the structure of meiotic chromosomes leads to an understanding of the nature of meiosis. The first prophase of meiosis, in which the crossing-over takes place, should be paid particular attention.

2. Lampbrush chromosomes

Oocytes – female reproductive cells are formed in oogenesis in ovaries. After the mitoses cease, the oogonia (primary sex cells) become first-order or primary oocytes. Next, they quickly enter the S phase, the preleptotene during which DNA is replicated for the last time. The oocyte grows during the meiotic prophase and, in most animals, stops expanding in the metaphase of the first meiotic division. Most of this cell growth takes place in the diplotene stage. In that stage, diplotene chromosomes of some vertebrates, e.g. birds assume the form of lampbrush chromosomes and generate thousands of loops along their axis, interpreted as sites of transcriptional activity (Macgregor & Varley, 1988; Morgan, 2002).

Lampbrush chromosomes still need to be fully explored. It is still unknown how they form out of the small mitotic structures, nor how they function in the oocyte. What is known is that they are intermediary structures present in the first meiotic division in the prolonged diplotene stage. They originate from a small telophase form at the end of the last oogonium mitosis. As they enter the diplotene stage of the first meiotic division, they undergo rapid decondensation that generates very large chromosome structures (Macgregor & Varley, 1988; Schmid et al., 2005).

Lampbrush chromosomes were discovered in 1882 by Flemming who observed salamander egg cells (*Ambystoma mexicanum*). Ten years on, LBCs were identified in shark egg cells and described by Rückert. It was Rückert who introduced the term "lampbrush chromosome" into biological nomenclature. The chromosomes take their name from 19th century brushes for cleaning street lamps to which Rückert likened them. The modern version of the item are bottle or test-tube brushes (Fig. 1) (Callan, 1986; Macgregor, 1977, 1980, 1987; Macgregor & Varley, 1988).

Lampbrush chromosomes are intermediate structures present during the first meiotic division. In the prolonged diplotene stage, they undergo decondensation that produces very large chromosomal structures. LBC length ranges (depending on the species) from 400 to 800 μm, which makes them up to 30 times larger than their mitotic counterparts (Callan, 1986; Callan et al., 1987; Rodionov, 1996). The basic profile of LBCs is performed with a 20x zoom of the microscope. In the case of avian mitotic chromosomes, a 20x zoom only makes it possible to identify the metaphase plate, not always enabling the determination of the number of chromosomes.

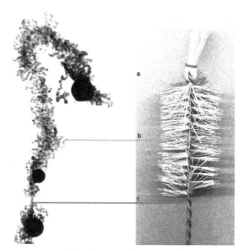

Fig. 1. A lampbrush chromosome and the "original item". The arrows indicate analogous structures; a- telomeric loop, b- side loops, c- a chromatid without loops (Katarzyna Andraszek).

Figure 2 shows a 20-fold microscopic magnification of the metaphase plate (a) compared in size with (b) 100-fold magnification of the second-pair mitotic chromosome) a 20-fold magnification of the second lampbrush bivalent (c). The arrow shows the second-pair mitotic chromosome on the metaphase plate.

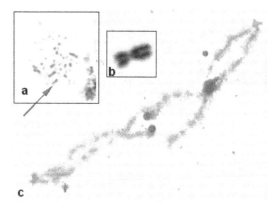

Fig. 2. A comparison of the size of LBC and mitotic chromosomes (Katarzyna Andraszek).

In the early prophase, a lampbrush chromosome is a bivalent that consists of two conjugating homologues ultimately becoming a tetrad. The axis of each of the homologue chromosomes is constituted by sister chromatids. Each chromatid is composed of alternately positioned regions of condensed inactive chromatin (chromomeres visible as dark irregular structures and also observed in the interphase nucleus) and side loops of decondensed chromatin. In the homologous sections of the bivalent, chromatin is condensed (spirally twisted) or decondensed in the form of side loops – two per each chromosome and four at the level of the bivalent. The loop constitutes a part of the chromosome axis. It is both extensible and contractible. The contractibility of the loop results in the contraction and dilation of the chrommomere (Angelier et al., 1984, 1990, 1996; Chelysheva et al., 1990; Macgregor, 1987; Morgan, 2002).

The use of a 100x zoom to analyse LBC structure made it possible to observe chromomeres, chiasmata and sister chromatids of each bivalent homologue. The identical zoom used for the analysis of avian mitotic chromosomes enables only the identification of their morphological structure in relation to the first couple of macrochromosome pairs. Figure 3 shows a 20-fold magnification of the second goose bivalent (a) and its distinctive structures visible with a 100x zoom (b, f – telomeres, c – centromere, d – chiasm, e - sister chromatids). In the case of the structural analysis of male meiotic chromosomes it is not possible to observe these crucial meiotic cytogenetic features.

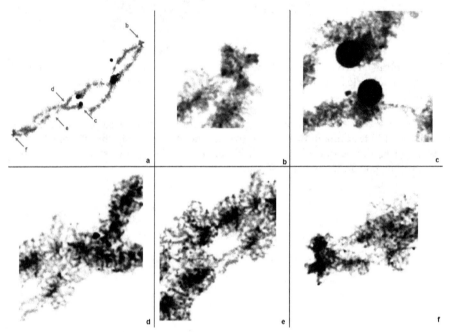

Fig. 3. The second goose LBC with a magnification of its distinctive structures (Katarzyna Andraszek).

As a tool, lampbrush chromosomes were introduced into poultry cytogenetics by Kropotova and Gaginskaya (1984), and by Hutchison (1987). The former authors support a thesis that chromosomes provide valuable information on bird gene expression, and are irreplaceable in cytogenetic research on animals with small genomes in which a large number of small-sized mitotic chromosomes makes it impossible for scientists to carry out microchromosome analysis. Just like in the case of banded patterns of mitotic chromosomes, LBCs are characterised by a special arrangement of active and inactive chromomeres visible as a pattern of side loops and regions without loops. In the Second Report on Chicken Genes and Chromosomes LBCs have been recognised as a new model in avian cytogenetics (Schmid et al., 2005).

Lampbrush chromosomes represent a new model in avian cytogenetics and are increasingly more often used in poultry chromosome analyses. Additionally, lampbrush chromosomes are considered as model structures in the study of transcription regulation. Changes in transcription activity are reflected as modifications of LBC morphological structure and are associated with physiological processes of the organism. The transcription activity analysis is carried out according to the concept assuming that it takes place in LBC side loops.

The aim of the present study (Andraszek et al., 2009; Andraszek & Smalec, 2011) was to compare the structure of the first five lampbrush macrochromosomes and ZW sex lampbrush bivalents, sampled from the oocytes of geese prior to and after the reproductive period and compare the transcription activity of lampbrush chromosomes and the G band pattern of corresponding mitotic chromosomes of the European domestic goose Anser anser.

The pre-reproduction bivalents were marked with lowercase "a", the post-reproduction ones with lowercase "b". The marker structures of the bivalents were successively numbered. The structure of the lampbrush chromosomes was analysed paying special attention to the comparison of the transcription-active parts and the GTG pattern on the corresponding mitotic chromosomes. The following marker structures were identified in the LBCs under analysis: GLLs – giant lumpy loops, MLs – marker loops, DBLs – distal boundary loops, PBLs – proximal boundary loops, TLs – telomeric loops, TBLs – telomeric bow-like loops, TLLs - telomeric lumpy loops, DBs – double bridges, Chs – chiasmata, PBs – protein bodies.

The respective bivalents sampled prior to and after reproduction have similar sizes but differ in morphological structure. The lampbrush chromosomes sampled after reproduction have reduced side loops – sites of intensive transcription activity. On the other hand, inactive chromomeres become prominent in the chromosomes. Marker loops are those structures that are degraded last after the end of reproduction. Consequently, they are used as the basis for identifying particular bivalents at different stages of transcriptional activity of the cell. The dark blocks correspond to the location of transcription active regions.

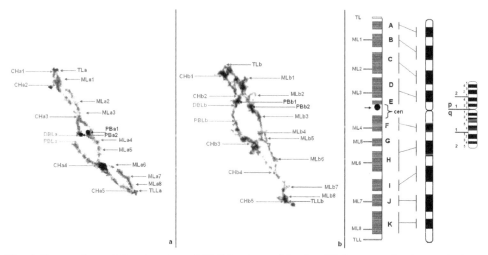

Fig. 4. Comparison of the structure of LBC 2 prior to (a), after (b) the reproductive period and graphic comparison of transcription activity of the second LBC and G bands on the second mitotic chromosome (c). Arrows indicate marker structures of bivalents.

The comparison of the location of regions with and without loops of the analysed lampbrush chromosomes with the GTG pattern of the corresponding mitotic chromosomes revealed that the arrangement of regions with side loops on LBCs which are transcription active corresponded to the GTG pattern on the mitotic chromosomes.

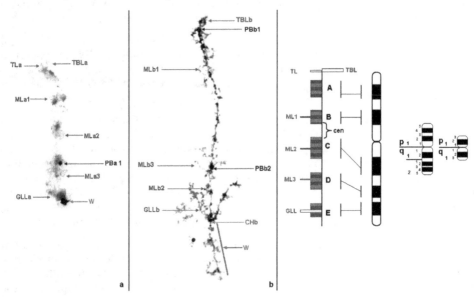

Fig. 5. Comparison of the structure of sex LBC prior to (a), after (b) the reproductive period and graphic comparison of transcription activity of the sex LBC and G bands on the sex mitotic chromosome (c). Arrows indicate marker structures of bivalents.

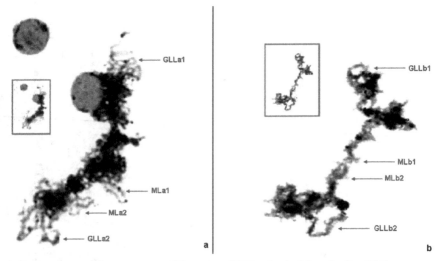

Fig. 6. Comparison of the structure of the micro-LBC prior to (a) and after (b) the reproductive period. Arrows indicate marker structures of bivalents.

Particular interest in recent years has been devoted to possibilities of using lampbrush chromosomes in genome mapping. This strategy can combine chromosome marker mapping and physical gene mapping using the in situ hybridisation technique with genetic maps constructed on the basis of chiasm incidence in the analysed bivalents. Equally important is also the possibility of using lampbrush chromosomes in analyses of the interaction of genes with other cellular structures. Particularly promising seem to be the possibilities of using lampbrush chromosomes in the mapping of avian genomes. This strategy can combine chromosome marker mapping and physical gene mapping using the FISH technique with genetic maps constructed on the basis of chiasm incidence in the analysed bivalents.

3. Meiosis during spermatogenesis

The cytogenetic analysis of breeding animals is problematic in the case of birds. The reason for the stalemate in research are the specific characteristics of the avian karyotype. A typical avian karyotype consists of several or a dozen or so macrochromosomes and about sixty microchromosomes. (Christidis, 1989; 1990). The only systematic description of avian chromosomes is the karyotype of the domestic hen (Smith et al., 2000).

A high number of chromosomes and the presence of microchromosomes in the bird karyotype made cytogeneticists look for other sources of information on the chromosomes. An important material for the analysis of the avian karyotype are meiotic chromosomes. The earliest studies of meiotic chromosome structure were conducted in the United States. The first investigations provided information on the huge variability of meiotic chromosomes, even within one genus. Among Batrachoseps salamanders, some species were unique in having a specific stage of chromosome dispersal between pachytene and diplotene, whereas no such phase was observed in other species (Macgregor & Varley, 1988).

Meiosis is usually observed in males, as spermatocytes are relatively small, numerous and easily accessible. The observation of meiotic chromosomes in females is impeded by the large size of the egg cell, technical difficulty in acquiring chromosomes from the egg cell and the fact that, compared with spermatocytes, oocytes are so few. Meiotic chromosomes are most often analysed in pachytene and diplotene cells during the first meiotic division. Cells typical of the 2nd meiotic division are few. Therefore, colchicine and vinblastine sulphate are applied to increase the number of the cells (Fechheimer, 1990).

In contrast to mitotic chromosome analyses, the number of studies of avian meiotic chromosomes is limited. Miller was the first to analyse meiosis in Gallus domesticus in 1938. He used primary spermatocytes of cocks as the experimental material. He identified 39 or 40 bivalents in metaphase I (Ford & Woollam, 1964). Similar experiments were performed by Ohno (1961). He identified 39 bivalents in Gallus. Moreover, he concluded that micro- and macrochromosomes follow identical behaviour during meiosis. The abovementioned studies concerned only diakinesis and the pre-meiotic phase and were conducted on a small number of cells sampled from several animals. Comprehensive information on meiotic chromosomes and the process of meiosis in hens was provided by Pollock & Fechheimer (1978). They analysed chromosomes at every stage of meiosis. Additionally, the number of chiasmata in the bivalents was determined. A separate group of studies of meiotic chromosomes in Gallus is constituted by analyses of synaptonemal complex structure both in spermatocytes (Kaelbing and Fechheimer, 1983, 1985) and oocytes (Solari, 1977; Rahn & Solari, 1986; Solari et al., 1988).

A paper by Andraszek and Smalec (2008) presents the course of meiosis in goose spermatocytes, with particular emphasis on the first meiotic prophase, the stage at which crossing over takes place, a process that guarantees recombination variability of organisms.

The results provided in the publication represent the only available study of avian meiosis in the relevant literature. They can be treated not only as practical reference data on the cytogenetics and biology of bird reproduction but also as didactic material.

Figure 7a shows the early prophase of the first meiotic division - leptotene. At this stage, chromosomes have the form of thin and long chromatin threads within which the chromatids cannot be distinguished. It is impossible to distinguish particular chromosomes which resemble an entangled wool-ball. In Figure 7b, early prophase chromosomes are also visible. Gradual chromatin condensation that occurs during the entire prophase causes the chromosome looping to become looser offering the possibility to distinguish particular chromatin threads. At this stage it is still impossible to distinguish particular chromosomes.

Figure 7c shows an image of late-zygotene bivalents. Due to progressive chromatin condensation particular chromosomes become separated. They are still long, while the chromatids remain indistinguishable. At this stage in the prophase, it is already possible to discern macro- and microchromosomes. However, it still remains impossible to identify particular homologous chromosomes, as they conjugate over their entire length. At this stage, synaptonemal complexes are observable only with an electron microscope. Figure 7d shows late-pachytene bivalents. Due to progressive chromatin condensation, particular bivalents are already distinguishable. It is possible to discern particular bivalents. Meiosis is not a strictly synchronous process. In some cells it is still impossible to recognise the homologues of the bivalent, since they conjugate over their entire length. Arrows in Figure 7e show the initiation sites of synaptonemal complex degradation . Figure 7f shows typical early-diplotene chromosomes. The cell is preparing for crossing over. The bivalents begin to divide into chromatids. At some points in the chromosome it is possible to observe chiasmata (indicated with arrows). The next Figure 7g depicts meiotic chromosomes in early diakinesis. Arrows indicate the prominent chiasmata, discernible due to gradual chromosome condensation. At this stage, the chromosomes are already after recombination. During the entire prophase, the nucleoli disintegrate. However, they remain observable until the end of the first meiotic prophase. Different numbers of bivalent-specific nucleoli (indicated with arrows) (Figure 7h) of different sizes (figure 7i) were observed in the analysed preparations. Figure 7j shows an image of a cell in late diakinesis, after the terminalisation of chiasmata. The chromosomes, with discernible micro- and macrochromosomes, contain condensed chromatin. A cell with such organisation enters the metaphase of the first meiotic division.

The work did not attempt to profile the cells in the remaining meiotic stages. Cells that are observed during the second meiotic division are the rarest category of reproductive cells in cytogenetic analysis. Even if they are visible under the microscope, the quality of definition is not satisfactory enough to enable description. Apart from that, the authors intended to focus on the prophase of the first meiotic division, the stage at which the recombination of the genetic material occurs.

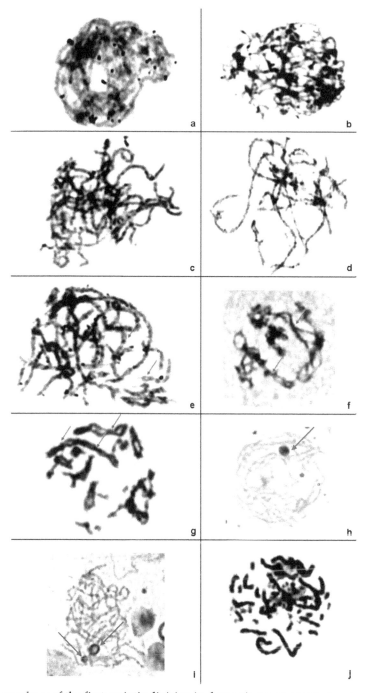

Fig. 7. The prophase of the first meiotic division in domestic goose spermatocytes.

4. Synaptonemal complex

A structure that is inextricably associated with the meiotic division of the cell is the synaptonemal complex (SC) - a proteinic structure which binds homologous chromosomes during the prophase of the 1st meiotic division and ensures correct genetic recombination. Apart form mutative variability, recombination variability is the reason for the huge diversity of organisms. Recombination occurs within the two homologous chromosomes that mutually exchange fragments of chromatids. This produces unique combinations of alleles of maternal and paternal origin in the genome. Due to crossing-over and DNA content reduction to the level of 1C, diploid organisms generate gametes, thus being able to create a new organism, different from the parental one.

An immense role in this process is played by the synaptonemal complex that "clasps" together two parallel homologous chromosomes and enables their conjugation (synapsis) (Turner et al., 2004). The synaptonemal complex was first observed more than fifty years ago in the spermatocytes of crayfish (Moses, 1956) and, subsequently, the dove, cat and man (Fawcett, 1956). The name of this unique nuclear structure was coined slightly later on. This meiotic structure is evolutively conservative in the bulk of sexually reproducing eukaryota, including the nonnucleated Protozoa, Fungi and algae, as well as vertebrates (Marec, 1996; Penkina et al., 2001)

A fully formed synaptonemal complex is situated between two prophase homologous chromosomes, binding them along their entire lengths into a pair that constitutes the bivalent. This is a three-tier proteinic structure that consists of two lateral elements (LE) and a central element (CE) located between them. Between the lateral elements there are recombination nodules (RN). They are ellipsoid, highly electron-absorbing protein complexes which are present only in euchromatinic regions of the chromosome (Holm & Rasmussen, 1980; Schmekel & Daneholt, 1998). At the centre of the complex it is possible to notice the so-called ladder structure produced by the linking of lateral elements with the central element using microfilaments (TFs) (Marec, 1996; Penkina et al., 2002).

The molecular structure of the synaptonemal complex can only be analysed using an electron or scanning microscope. An optical microscope, even with a high resolution and zoom, makes it possible to make out only the bivalent structure with clear-cut synaptic chromosomes. The available sources equate bivalent presence with the synaptonemal complex due to the fact that bivalent existence directly results from the presence of the synaptonemal complex. However, the bivalent and the synaptonemal complex are two separate structures. Figure shows prophase, meiotic chromosomes of a European domestic goose male (a), the ladder-like structure of the bivalents indicates that the synaptonemal complexes have fully developed (ringed in the photo). Alongside: a schematic structure of the synaptonemal complex (b).

The synaptonemal complex begins to form during the first meiotic prophase. At that time chromatin organisation undergoes dramatic changes. Starting with leptotene, chromatid DNA (present as chromatin loops) begins to connect to proteinic elements that constitute the matrix for emergent LEs. Homologous chromosomes have to be positioned in the distance of approx. 300nm for synapsis to occur (Marec, 1996). In zygotene, the homologues start to approach each other and connect using SC elements (Penkina et al., 2002).

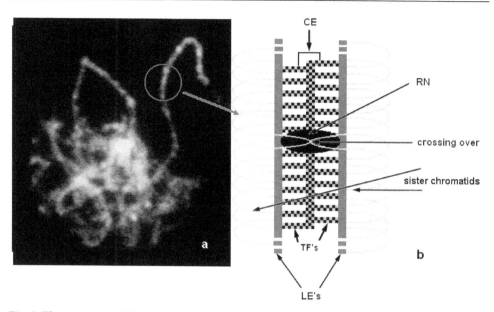

Fig. 8. The structure of the synaptonemal complex, Anser anser meiotic chromosomes, DAPI staining - (a); an outline of the synaptonemal complex – (b).

Chromatin loops depart radially from SC lateral elements. Most of the SC-bound chromatin is inactive – it is not transcribed (Marec, 1996). During the early prophase, LEs appear as proteinic axes (axial elements), each connected with two sister chromatids. Lateral elements are always built out of newly synthetised proteins, never as a result of the reorganisation of already existing components (Heyting et al., 1989). At first, LEs are visible as single, short fragments entangled in chromatin. Next, they become anchored to the inner side of the nuclear membrane with the aid of telomeres. The anchoring site of SCs is always situated in the region of the nucleus that is located opposite to the nucleolus, close to the diplosome. Lateral elements become continuous in mid-zygotene, when first complete bivalents appear and telomeres enter the distinctive bouquet stage at SC-anchoring sites of the nuclear membrane. LEs then stretch from one telomere to another in each chromosome. After LE formation, the CE begins to appear, connecting the homologues in a "zipper" fashion over their entire length, except for the sites at which the process has been inhibited by interlocking (Marec, 1996).

In early pachytene, all the homologues are in full-blown synapsis. Fragments of chromosomes that form the nucleolar organizer region are the latest to attain complete synapsis (Rasmussen, 1986; Marec, 1996). The formation of the complex in yeast and higher plants is initiated at many points in the bivalent, whereas in animals it always starts from the telomeric regions and and progresses lengthwise. Interstitial synapsis also occurs, if homologue mobility is impeded due to interengaging with other homologues – interlocking (Rasmussen, 1986; Penkina et al., 2002). SC formation is initiated in subterminal regions of the chromosome and progresses towards the nearest telomere. The above observations show that chromosome conjugation is caused by the existence of two identification sites in each chromosome, not by absolute homology of chromosome regions. The SC retains its structure until late diplotene. The first to disintegrate are CEs. Next, LEs crack along their longitudinal axis. They are decondensed in a number of stages and the SC is removed from the bivalents (Penkina et al., 2002).

4.1 Heterochromosomal synapsis

In the cells of homogametic organisms sex chromosome pairing does not differ from autosome pairing. In species whose sex is determined by the XY or ZW pair a considerable morphological and genetic variability of these chromosomes is observed. What is problematic is the length and the gene composition of the chromosomes. In the majority of such cases chromosome behaviour during conjugation seems to be forced and slightly unnatural. Despite the abovementioned impediments, the process runs correctly, which testifies to a huge adaptive potential and dynamics of the complex (Marec, 1996; Page et al., 2006).

All of the karyologically studied birds (about 10% of the living species) have a heterogametic system of sex determination in females in the form of the ZW pair. Avian sex chromosomes are rich in euchromatin, with the exception of centromeric regions and the short arm of the W chromosome. The W chromosome is often metacentric and entirely consists of heterochromatin. During pachytene, the axes of the Z and W chromosomes form a bivalent bound by a synaptonemal complex. The length of the SC corresponds to the length of the W chromosome axis. Next, from late to mid-pachytene, the unpaired section of the Z arm shortens into a streamer-like structure so as to assume the length of fully developed SC lateral elements. In this way, the Z arm twists and forms a loop around the straight W arm. The size of this loop corresponds to the size of the non-homologous pairing region in this bivalent (Pigozzi & Solari, 1999).

The synaptonemal complex in higher organisms begins to form starting from the telomeric regions. It might seem that also in birds shorter microchromosomes would be the first to form synapsis. In preparations sampled from one-day-old Anser anser goslings it was observed that macrobivalents were the first to accomplish synapsis (Andraszek et al., 2008). Figure9a shows a cell with the developed first macrobivalent (closed arrow). The dark structures in the subproximal area of the bivalents and the dark structures on the background of chromatin are kinetochores which become intensely hued after silver nitrate staining (open arrows). Kinetochores in the subproximal regions of macrochromosomes (open arrows) evidence the submetacentric forms of the first two goose macrobivalents (Figure 9b). Figure 9c shows a cell with developed synaptonemal complexes within the first four macrobivalents and the ZW univalent, as well as within the acrocentric bivalents differing successively in size.

The phenomenon provides an explanation to the specific character of the bird genome which, unlike in mammals, possesses a marked number of telomeric sequences in the interstitial parts of chromosome arms (Solovei et al., 1994; Nanda et al., 2002). The occurrence of interstitially located telomeric sequences on macrochromosomes is connected with an increase in the number of places where the formation of SCs is initiated, which explains why synapsis is achieved quicker. Santos et al. (1993) observed that in submetacentric bivalents synaptonemal complex formation begins from subtelomeric regions of chromosomes. In acrocentric bivalents there exist two potential initiation sites which are in the distal and proximal parts of the q arm. Short p arms pair much later.

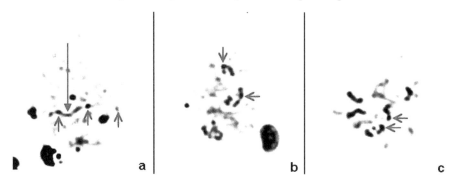

Fig. 9. Chromatin organisation in one-day-old gosling cells; isolation – Counce & Meyer (1973), staining – Howell & Black (1980)

In geese the largest macrochromosomes are submetacentric whereas the largest microchromosomes are acrocentric, so the theory of Santos et al. (1993) and the presence of interstitial telomeric sequences in chromosomes may explain why it is macrochromosomes and not microchromosomes that are the first to undergo synapsis. Moreover, compared with macrochromosomes, microchromosomes are characterised by a lower adenine-thymine content which is accompanied by a higher guanine-cytosine content (Fillon et al., 1998; Gregory, 2002). G-C-rich sequences constitute about 60% of Zyg DNA which is responsible for the linkage with SC. Zyg DNA undergoes replication not during the pre-meiotic S-phase but as late as in zygotene (Marec, 1996) so synapsis formation is delayed in those places.

It remains unexplained whether non-simultaneous synapsis is intended or coincidental, a consequence of the occurrence of interstitial telomeric sequences or Zyg DNA sequences, considering that most genes are concentrated in delayed microchromosomes (Fillon et al., 1998; Gregory, 2002). Perhaps, this is connected with the loop repair mechanism of chromosomes stuck due to interlocking. Later synapsis makes it possible for the repair mechanism to thoroughly verify whether the bivalent structure is correct, which is a way of preventing the loss of a valuable chromosome fragment. Unfortunately, this hypothesis cannot be testes as, apart from birds, no animal group has micro- and macrochromosomes in such a form.

The distinctive morphology of the ZW pair, in the form of a univalent in which chromosome W is bound to the distal part of chromosome Z, is also a consequence of synaptonemal complex formation starting from subtelomeric sequences. Moreover, as all the chromosomes start to connect and form a bivalent from telomeric sequences, the risk of losing their distal endings is reduced to a minimum. This would not be possible, if the complex formation proceeded from the centromere towards the chromosomal endings. One can think that the role of the SC is to stabilise the bivalent structure, in parallel to the role of telomeres that stabilise the chromosome structure.

The application of conventional techniques of synaptonemal complex identification and preparation staining does not make it possible to identify and analyse syn-aptonemal complex molecular structure. The presence of SCs on the preparations obtained from one-day-old goslings can be inferred from the presence of macrobivalents and darkly stained kinetochores and subtelomeric regions. In preparations obtained from 17-week-old ganders, the presence of the complexes was once again deduced from the presence of a complete set of bivalents as well as the number of kinetochores typical of the haploid set of chromosomes.

Figure 10a shows a cell with all the bivalents already visible. This is evidenced in the number of kinetochores. In the macrobivalents, kinetochores in the subproximal regions of the bivalents (open arrow) and dark-hued subtelomeric regions (closed arrow) are identifiable Figure 10b shows all the bivalents fully developed. The macrobivalents are long and well identifiable. It is possible to distinguish kinetochores and telomeric regions within them. It is also possible to observe the different lengths of the bivalents, ranging from the longest first one (closed arrow) to the very short acrocentric microbivalents (open arrow).

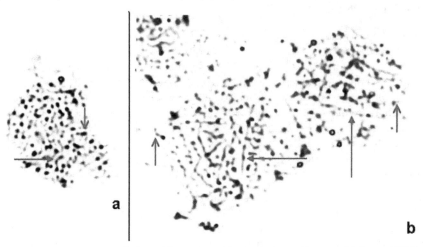

Fig. 10. Bivalent structure in 17-day-old gander cells; isolation – Counce & Meyer (1973), staining – Howell & Black (1980)

The standard technique for synaptonemal complex identification is the Counce and Meyer method (1973). The preparations for which it was used are presented above, in photograph 11a and 11b. On the other hand, promising results were obtained using the technique of meiotic chromosome isolation described by Pollock and Fehcheimer (1978), followed by silver nitrate staining and DAPI fluorochrome staining tentatively used in the experiment. After the application of the above techniques the homologues and bivalents were easily discernible. Additionally, the distinctive, ladder-like structure of synaptonemal complexes resulting from a different protein composition of the lateral elements was observed. Moreover, in the DAPI-stained preparations unique structures were identified during the basic cytogenetic analysis - two parallel homologous chromosomes just before synaptic union (Andraszek et al., 2008).

Figure 11a shows a cell with developed synaptonemal complexes within the bivalents. The structure of the complex is well visible both in the macrobivalents (closed arrow) and microbivalents (open arrow). Figure 11b shows a cell at the onset of synaptonemal complex degradation. It is possible to distinguish particular homologues of the bivalent (arrows). The distinctive synaptonemal complex structure can still be observed as alternating dark and light chromatin regions (circled) similar to the high-fidelity banding pattern of mitotic chromosomes. Such cells are typical of late pachytene and the onset of diplotene. Figure 11c shows a cell in zygotene. The homologues of the bivalents are already connected by synapsis (open arrow). In turn, the closed arrows indicate the bivalent homologues before the complete stabilisation of the SC. In Figure 11d , the bivalents have the typically synaptic ladder-like structure and prominent light fluorescent telomeres and kinetochores (open arrow) over their entire length . It is also possible to see the onset of synaptonemal complex disintegration (closed arrow) and two clear bivalent homologues after the break-up of synapsis.

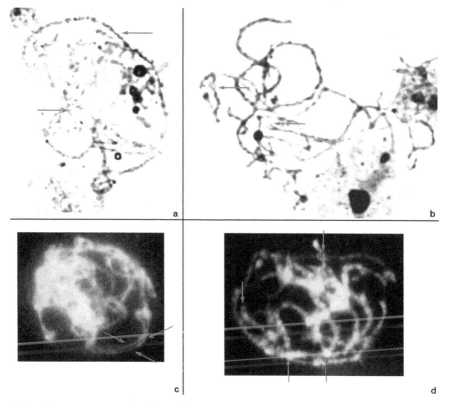

Fig. 11. Bivalent structure in 17-day-old gander cells; isolation – Pollock & Fehcheimer (1978), staining a, b – Howell & Black, c, d - Schweizer et al. (1978).

Paired or unpaired chromosome regions may indicate the chromosomal position of such marker structures as centromeres, nucleolar organizers, nucleoli, telomeres or even heterochromatin regions.

Such preparations may be very useful for analysing prophase chromosomes, and generating the so-called SC karyotypes. The karyotypes would make it possible to observe pairing initiations, sequences participating in pairing, the frequency of recombination occurrence, as well as potential meiotic chromosome structure anomalies. What follows is the fact that the complex lateral elements connect to DNA in strictly determined sequences. Thus, by convention, the ladder-like SC structure can be treated as a bivalent banding pattern indicating, similarly to banded staining, the location of specific sequences in the chromosome.

5. Nucleolus

The nucleolus is the largest and best known functional constituent of the cell nucleus. It is formed by original products of ribosomal RNA genes contained in the nucleolar organiser, related proteins and various enzymes, including RNA polymerase, RNA methylase and RNA endonuclease (Shaw & Jordan, 1995; Scheer & Hock, 1999; Hernandez-Verdun, 2006). The nucleolus is the site of synthesis and maturation of ribosomal RNA (rRNA) molecules. The molecules also bond with proteins in the nucleolus. rRNA genes are situated in particular chromosomes, in nucleolar organizer regions (NORs) that participate in the formation of nucleoli (Olson, 2004; Raška et al., 2004, 2006; Lam et al., 2005; Prieto & McStay, 2005; Derenzini et al., 2006). Nucleoli are present in the nuclei of almost all eukaryotic cells since they contain elementary metabolic genes, with the exception of spermatozoa and mature avian erythrocytes (Kłyszejko-Stefanowicz, 2002; Raška et al., 2006).

The weight of nucleoli is higher than that of the nucleoplasm in which the nucleoli are suspended. On account of its distinctive density, compact structure and low water content (10%) the nucleolus is an organelle that can be readily distinguished. Due to its characteristics, following cell nucleus fragmentation, the nucleolus remains intact in the saline solution even after the destruction of most nuclear structures, thereby allowing its isolation through centrifugation. The isolated nucleolus is identical to the one present in the nucleus of a living cell, and even retains its transcriptional activity in some cases (Olson, 2004; Hernandez-Verdun, 2006; Raška et al., 2006).

The number of nucleoli in a cell nucleus is determined by the number of active nucleolar organiser regions (NORs). It may be equal to the number of those NOR-chromosomes. Yet, normally, it is lower. This can be explained either with the fusion of nucleoli in the interphase nucleus or the suppression of activity of certain rDNA loci (Kłyszejko-Stefanowicz, 2002; Raška et al., 2006).

Nucleoli are very dynamic structures. This may be reflected in their cyclical disappearance during mitosis and reappearance at its end (Scheer & Benavente, 1990). The nucleolus disappears during cytokinesis and is reproduced in the reconstructed nuclei as a result of NOR activity (Kłyszejko-Stefanowicz, 2002; Olson, 2004) Nucleolar material appears between the chromosomes during the reconstruction of the NOR-associated telophase nucleus. Next, rRNA synthesis is resumed causing the nucleoli to become more visible. During the interphase, the nucleolus is spherical in shape. In the prophase, when the chromosomes become visible, it is evident that the nucleoli are associated with particular nucleolus organising chromosomes (Raška et al., 2006)

Both nucleoli and nucleolar organiser regions have chemical affinity for heavy metals and can be identified through staining with silver nitrate in a protective colloidal solution of formalin or gelatine (Howell & Black, 1980). Since nucleolar organiser regions (NORs) determine the structure of nucleoli, an alternative source of information on the activity of avian rRNA-encoding genes can be found in the analysis of the numbers and sizes of nucleoli in the prophase of the first meiotic division. Throughout the prophase, nucleoli are not degraded. As opposed to mitotic NORs, they are large structures (Andraszek & Smalec, 2007; Andraszek et al., 2009a; Andraszek et al., 2010a, 2010b).

The nucleoli disintegrate throughout the entire prophase of the first meiotic division. The way in which they decay is probably typical of particular vertebrate groups, this being possibly a species-specific characteristic. In the spermatocytes of domestic cattle, the nucleoli gradually become fragmented and "disintegrate" into tiny structures whose number corresponds with the number of NOR regions (Andraszek, unpublished). The studies which analysed the number and size of nucleoli in avian spermatocytes (Andraszek & Smalec, 2007; Andraszek et al., 2010b) reported a different mechanism of disappearance of nucleoli. In birds, chromatin reorganisation during the prophase of the first meiotic division and the related change in the cell nucleus size is correlated with decreasing sizes of the nucleoli. At the beginning of the prophase, in the early leptotene, the nucleoli are visible as large oval structures. In turn, at the end of the prophase they are observed as tiny points associated with specific bivalents. The figure 12 shows the different sizes of nucleoli at the beginning and end of the first meiotic prophase.

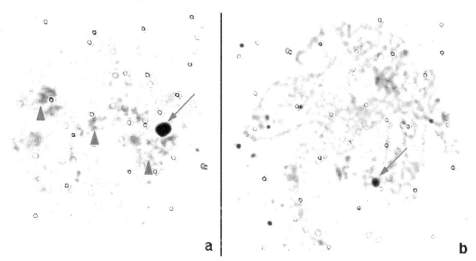

Fig. 12. Meiotic chromosomes of quail – early prophase (a), late prophase (b). The nucleoli indicated with arrows, kinetochores – arrow points.

Another distinctive characteristic of nucleoli in avian cells are the variations of their sizes unrelated with the prophase stage. Different sizes of nucleoli can be observed in cells that are at the same stage of meiosis. Figure 13 shows different sizes of nucleoli in goose (a), chicken (b) and quail spermatocytes (c).

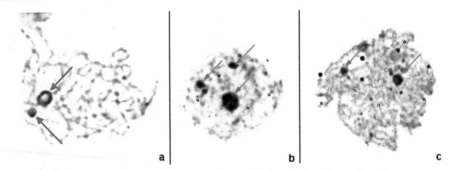

Fig. 13. Variability of the sizes of nucleoli in cells. The nucleoli indicated with arrows.

Avian cells also have variable numbers of nucleoli. This variability occurs at the individual, cellular and interindividual level. The number of nucleoli in the spermatocytes of geese and hens ranges from 1 to 4, and from 1 to 2 in quail cells. Next figures shows different numbers of nucleoli in goose (figure 14), chicken (figure 15) and quail spermatocytes (figure 16)

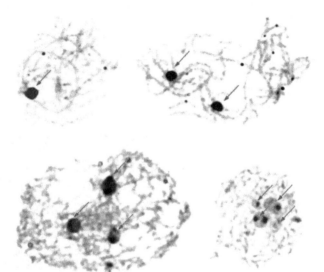

Fig. 14. Variability of the number of nucleoli in goose cells. The nucleoli indicated with arrows.

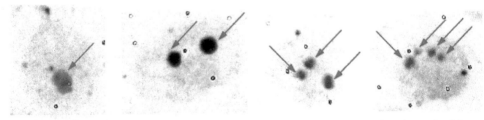

Fig. 15. Variability of the number of nucleoli in chicken cells. The nucleoli indicated with arrows.

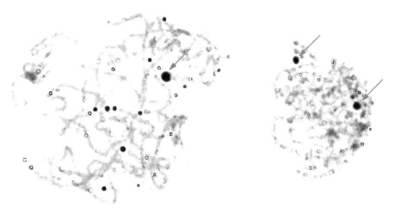

Fig. 16. Variability of the number of nucleoli in quail cells. The nucleoli indicated with arrows.

Apart from its functions being directly connected with ribosome biogenesis, the nucleolus is associated or involved in other cellular processes (Pederson, 1998; Santoro & Grummt, 2001; Olson et al., 2002; Gerbi et al., 2003; Raška et al., 2006). At present, it is not possible to determine whether these unconventional roles are the main functions of the nucleolus or adaptations of individual species or species groups. As regards the functions discussed further on, there is now a rich and still expanding body of literature available that deals with the relationship between the nucleolus and viral infections (Hiscox, 2002, 2003; Olson et al., 2002), including DNA, RNA and retroviruses.

Nucleolar morphology was one of the key criteria of neoplasm classification. The morphometric parameters of nucleoli are: the number, size and distance from the nuclear membrane (Nafe & Schlote, 2004; Smetana et al., 2005, 2006; Raška et al., 2006). Taking advantage of previous observations of Montgomery (1895), biologists dealing with neoplasms quickly tracked down the connection between AgNOR stains and cell proliferation (Derenzini et al., 1990; 2006; Raška et al., 2006).

Most studies during the last few years exploited the potential offered by the newly discovered nuclear oncogenes connected with the promotion and inhibition of tumours by cytogenetically diagnosed nucleolar mechanisms. c-Myc proteins (the product of the c-myc proto-oncogene) are located in the nucleolus and control rRNA synthesis (Oskarsson & Trumpp, 2005; Raška et al., 2006). It has also been shown that c-Myc is capable of controlling the activity of all the three polymerases in mammalian cells and coordinating the entire ribosome synthesis and cell growth (Arabi et al., 2005). These observations point at the crucial role of c-Myc in the development of promotional neoplastic actions via ribosome biogenesis control.

pRb (the protein of malignant retinoblastoma) and p53 proteins play a major role in the control of the cell cycle progress, as well as ensuring the correct development of daughter cells. These are oncosuppressive proteins, concentrated in the nucleolus (Ryan et al., 2001; Trere et al., 2004). The ARF/p16INK4a encoding gene is the second most common inactive human neoplastic gene (Ruas & Peters, 1998). ARF is situated at the nucleolus where it is associated with p53 (Kashuba et al., 2003). It has also been observed that there is a direct functional link between the nucleolus and p53 control (Ryan et al., 2001; Olson 2004). Most stress treatments activating p53 also cause the breakdown of nucleolar structure.

A considerable number of somatic cells do not have active telomerase. This puts a limit to the number of cell division cycles and suggests that telomerase is a factor in the processes of aging and carcinogenesis (Maser & DePinho, 2002; Raška et al., 2006). Telomerase has been found to be present in the nucleolus. Its function is connected with nucleolus activity control (Raška et al., 2006). Together with a class of snoRNA, telomerase locates RNA telomerase within the nucleolus (Lukowiak et al., 2001). hTERT – the reverse transcriptase of the catalytic subunit of human telomerase also has a nucleolar location (Khurts et al., 2004). Evidence of a functional relationship between telomerase and nucleoli has been provided by studies of human cell lines (Wong et al., 2002). In tumours and transmuted cell lines hTERT was eliminated from the nucleolus (Wong et al., 2002). The nucleolar phosphoproteinic nucleoid reacts with hTERT, the interaction being affected by RNA telomerase. This interaction is probably connected with dynamic telomerase anchoring. The TRF2 binding agent of the telomere is located in certain nucleolar modifications during the cell cycle (Khurts, 2004; Raška et al., 2006).

6. Conclusion

Meiosis guarantees the stability of the chromosome number in the consecutive generations of sexually reproducing organisms. Generally, meiosis is represented as a process during which the cell passes through totally different stages that correspond with particular structure and behaviour of chromosomes. The observation of the changes occurring in the structure of meiotic chromosomes leads to an understanding of the nature of meiosis. The first prophase of meiosis, in which the crossing-over takes place, should be paid particular attention. Moreover, meiotic chromosome research makes it possible to analyse the structure of nucleoli. In the meiotic prophase, they can be an alternative source of information on the activity of rRNA-encoding genes. In addition, synaptonemal complexes, which are extraordinary structures that guarantee recombination variability of organisms, are formed during meiosis. Summing up, meiotic chromosomes can not only be successfully used in applications in reproductive cytogenetics and biology but also as didactic aids.

7. References

Andraszek, K. & Smalec E. (2007). Number and size of nucleoli in the spermatocytes of European domestic goose (*Anser anser*). *Archiv für Geflügelkunde*, Vol. 71, No. 5, pp. (237-240), ISSN 0003-9098

Andraszek, K. & Smalec E. (2008). Spermatogenesis process as exemplified by meiosis in European domestic goose (*Anser anser*). *Archiv für Geflügelkunde*, Vol. 72, No. 3, pp. (110-115), ISSN 0003-9098

Andraszek, K. & Smalec E. (2011). Comparison of active transcription regions of lampbrush chromosomes with the mitotic chromosome G pattern in the

European domestic goose *Anser anser*. *Archiv Tierzucht*, Vol. 54, No. 1, pp. (69-82), ISSN 0003-9438

Andraszek, K.; Smalec, E. & Czyżewska, D. (2008). The structure of the synaptonemal complexes in meiocytes of European domestic goose (*Anser anser*). *Folia Biologica (Krakow)*, Vol. 56, No. 3-4, pp. (139-147), ISSN 0015-5497

Andraszek, K.; Horoszewicz, E. & Smalec, E. (2009a). Nucleolar organizer regions, satellite associations and nucleoli of goat cells (*Capra hircus*). *Archiv Tierzucht*, Vol. 52, No. 2, pp. (177-186), ISSN 0003-9438

Andraszek, K.; Smalec, E. & Tokarska, W. (2009b). Identification and structure of lampbrush sex bivalents prior to and after the reproduction period of the European domestic goose *Anser anser*. *Folia Biologica (Krakow)*, Vol. 57, No. 3-4, pp. (143-148), ISSN 0015-5497

Andraszek, K.; Danielewicz, A.; Smalec, E. & Kaproń, M. (2010a). Identification of the nucleoli in domestic horse spermatocytes - preliminary investigations. *Roczniki Naukowe PTZ*, Vol. 6, No. 4, pp. (13-21), ISSN 0003-9438

Andraszek, K.; Gryzińska, M.; Knaga, S.; Wójcik, E. & Smalec, E. (2010b).Number and size of nucleoli in the spermatocytes of chicken and Japanese quail. *Proceedings of 51*, Olsztyn, Poland, September 2010. ISBN

Angelier, N.; Paintraud, M.; Lavaud, A. & Lechaire, J.P. (1984). Scanning electron microscopy of amphibian lampbrush chromosomes. *Chromosoma*, Vol. 89, No. 4, pp. (243-253), ISSN 0009-5915

Angelier, N.; Bonnanfant-Jais, M.L.; Herberts, C.; Lautredou, N.; Moreau, N.; N'Da, E.; Penrad-Mobayed, M.; Rodriguez-Martin, M.L. & Sourrouille P. (1990). Chromosomes of amphibian oocytes as a model for gene expression: significance of lampbrush loops. *The International Journal of Developmental Biology*, Vol. 34, No. 1, pp. (69-80), ISSN 0214-6282

Angelier, N.; Penrad-Mobayed, M.; Billoud, B.; Bonnanfant-Jais, M.L. & Coumailleau P. (1996). What role might lampbrush chromosomes play in maternal gene expression? *The International Journal of Developmental Biology*, Vol. 40, No. 4, pp. (645-652), ISSN 0214-6282

Arabi, A.; Wu, S.; Ridderstrale, K.; Bierho, V.H.; Shiue, C.; Fatyol, K.; Fahlen, S.; Hydbring, P.; Soderberg, O.; Grummt I.; Larson, L.G. & Wright A.P. (2005): c-Myc associates with ribosomal DNA and activates RNA polymerase I transcription. *Nature Cell Biology*, Vol. 7, No. 3, pp. (303-310), ISSN 1465-7392

Callan, H.G. (1986). *Lampbrush chromosomes*. Springer-Verlag New York, Incorporated, ISBN - 10:0387164308, Berlin, Heidelberg, New York, Toronto

Callan H.G.; Gall J.G. & Berg C.A. (1987). The lampbrush chromosomes of *Xenopus laevis*: preparation, identification, and distribution of 5S DNA sequences. *Chromosoma*, Vol. 95, No. 4, pp. (236-250), ISSN 0009-5915

Chelysheva, L.A.; Solovei, I.V.; Rodionov, A.V.; Yakovlev, A.F. & Gaginskaya, E.R. (1990). Lampbrush chromosoms of the chicken: the cytological map of the macrobivalents. *Tsitologiia*, Vol. 32, No. 4, pp. (303-316), ISSN 0041-3771

Counce, S.J. & Meyer, G.F. (1973). Differentation of the synsptonemal complex and the kinetochore in Locusta spermatocytes studied by whole mount electron microscopy. *Chromosoma*, Vol. 44, No. 2, pp. (231-253), ISSN 0009-5915

Derenzini, M.; Pession, A. & Trere, D. (1990). Quantity of nucleolar silver-stained proteins is related to proliferating activity in cancer cells. *Laboratory Investigation*, Vol. 63, No. 1, pp. (137-140), ISSN 0023-6837

Derenzini, M.; Pasquinelli, G.; O'Donohue, M.F.; Ploton, D. & Thiry, M. (2006). Structural and functional organization of ribosoma, genes within the mammalian cell nucleolus. *The journal of histochemistry and cytochemistry*, Vol. 54, No. 2, pp. (131-145), ISSN 0022-1554

Fawcett D.W. (1956): The fine structure of chromosomes in the meiotic prophase of vertebrate spermatocytes. *The Journal of Biophysical and Biochemical Cytology*, Vol. 2, No. 4, pp. (403-406), ISSN 0095-9901

Fechheimer, N.S. (1990) Chromosomes of Chickens, In: *Domestic Animal Cytogenetics*, McFeely R.A., pp (169-207), Academic Press, ISBN 0120392348, New York

Fillon, V.; Morisson, M.; Zoorob, R.; Auffray, C.; Douaire, M.; Gellin, J. & Vignal, A. (1998). Identification of sixteen chicken microchromosomes by molecular markers using two color fluorescent in situ hybridization (FISH). *Chromosome Research*, Vol. 4, No. 6, pp. (307-313), ISSN 0967-3849

Ford, E.H.R. & Woollam, D.H.M. (1964). Testicular chromosomes of *Gallus domesticus*. *Chromosoma*, Vol. 15, pp. (568-578), ISSN 0009-5915

Gerbi, S.A.; Borovjagin, A.V. & Lange, T.S. (2003). The nucleolus: A site of ribonucleoprotein maturation. *Current Opinion in Cell Biology*, Vol. 15, No. 3, pp. (318-325), ISSN 0955-0674

Gregory, T.R. (2002): A bird's-eye view of the C-value enigma: genome size, cell size, and metabolic rate in the class aves. *Evolution*, Vol. 56, No. 1, pp. (121-130), ISSN 0014-3820

Hernandez-Verdun, D. (2006). Nucleolus: From structure to dynamics. *Histochemistry and Cell Biology*, Vol. 125, No. 1-2, pp. (127-137), ISSN 0948-6143

Heyting, C.A.; Dietrich, J.J.; Moens, P.B.; Dettemers, R.J.; Offenberg, H.H.; Redeker, E.J.W. & Vink, A.C.G. (1989). Synaptonemal complex proteins. *Genome*, Vol. 31, No. 1, pp. (81-87), ISSN 0831-2796

Hiscox, J.A. (2002). The nucleolus-a gateway to viral infection? *Archives of Virology*, Vol. 147, No. 6, pp. (1077–1089), ISSN 0304-8608

Hiscox, J.A. (2003). The interaction of animal cytoplasmic RNA viruses with the nucleus to facilitate replication. *Virus Research*, Vol. 95, No. 1-2, pp. (13-22), ISSN 0168-1702

Holm, P.B. & Rasmussen, S.W. (1980). Chromosome pairing, recombination nodules and chiasma formation in diploid Bombyx males. *Carlsberg Research Communications*, Vol. 45, No. 6, pp. (483-548), ISSN 0105-1938

Howell, W.M. & Black, D.A. (1980). Controlled silver-staining of nucleolus organizer regions with a protective colloidal developer a 1-step method. *Experientia*, Vol. 36, No. 8, pp. (1014-1015), ISSN 0014-4754

Hutchison, N. (1987). Lampbrush chromosomes of the chicken *Gallus domesticus*. *The Journal of Cell Biology*, Vol. 105, No. 4, pp. (1493-1500), ISSN 0021-9525

Kaelbling, M. & Fechheimer, N.S. (1983). Synaptonemal complexes and the chromosome complement of domestic fowl *Gallus domesticus*. *Cytogenetics and Cell Genetics*, Vol. 35, No. 2, pp. (87-92), ISSN 0301-0171

Kaelbling, M. & Fechheimer, N.S. (1985) Synaptonemal complex analysis of a pericentric inversion in chromosome 2 of domestic fowl *Gallus domesticus*. *Cytogenetics and Cell Genetics*, Vol. 39, No. 2, pp. (82-86), ISSN 0301-0171

Kashuba, E.; Mattsson, K.; Klein, G. & Szekely, L. (2003). p14ARF induces the relocation of HDM2 and p53 to extranucleolar sites that are targeted by PML bodies and proteasomes. *Molecular Cancer*, Vol. 2, No. 18, ISSN 1476-4598, < doi:10.1186/1476-4598-2-18>

Khurts, S.; Masutomi, K.; Delgermaa, L.; Arai, K.; Oishi, N.; Mizuno, H.; Hayashi, N.; Hahn, W.C. & Murakami, S. (2004). Nucleolin interacts with telomerase. *The Journal of Biological Chemistry*, Vol. 279, No. 49, pp. (51508-51515), ISSN 0021-9258

Kłyszejko-Stefanowicz, L. (2002). *Cytobiochemia*, (three edition), PWN, ISBN: 83-01-13824-6, Warsaw, Poland

Lam, Y.W.; Trinkle-Mulcahy, L. & Lamond, A.L. (2005). The nucleolus. *Journal of Cell Science*, Vol. 118, No. 7, pp. (1335-1337), ISSN 0021-9533

Lukowiak, A.A.; Narayanan, A.; Li, Z.H.; Terns, R.M. & Terns, M.P. (2001). The snoRNA domain of vertebrate telomerase RNA functions to localize the RNA within the nucleus. *RNA*, Vol. 7, No. 12, pp. (1833-1844), ISSN 1355-8382

Macgregor, H.C. (1977). Lampbrush chromosomes, In: *Chromatin and Chromosome Structure*, Eckhardt, R. A., and Hsueh-jei, Li., pp. (339-357), Academic Press, ISBN: 0124505503, New York and London

Macgregor, H.C. (1980). Recent developments in the study of lampbrush chromosomes. *Heredity*, Vol. 44, pp. (3-35), ISSN 0018-067X

Macgregor, H.C. (1987). Lampbrush chromosomes. *Journal of Cell Science*, Vol. 88, No. 1, pp. (7-9), ISSN 0021-9533

Macgregor, H.C, Varley J., „Working with Animal Chromosomes". John Wiley & Sons. London, New York, Brisbane, Toronto, Singapore, 1988.

Marec, F. (1996). Synaptonemal complexes in insects. *International Journal of Insect Morphology and Embryology*, Vol. 25, No. 3, pp. (205-233), ISSN 0020-7322

Maser, R.S. & DePinho, R.A. (2002). Keeping telomerase in its place. *Nature Medicine*, Vol. 8, No. 9, pp. (934-936), ISSN 1078-8956

Morgan, G.T. (2002). Lampbrush chromosomes and associated bodies: new insights into principles of nuclear structure and function. *Chromosome Research*, Vol. 10, No. 3, pp. (177-200), ISSN 0967-3849

Moses, M.J. (1956). Chromosomal structures in crayfish spermatocytes. *The Journal of Biophysical and Biochemical Cytology*, Vol. 2, No. 2, pp. (215-218), ISSN 0095-9901

Nafe, R. & Schlote, W. (2004). Histomorphometry of brain tumours. *Neuropathology and Applied Neurobiology*, Vol. 30, No. 4, pp. (315–328), ISSN 0305-1846

Nanda, I.; Schrama, D.; Feichtinger, W.; Haaf, T.; Schartl, M. & Schmid, M. (2002). Distribution of telomeric (TTAGGG)n sequences in avian chromosomes. *Chromosoma*, Vol. 111, No. 4, pp. (215-227), ISSN 0009-5915

Ohno, S. (1961). Sex chromosomes and microchromosomes of *Gallus domesticus*. *Chromosoma*, Vol. 11, pp. (484-498), ISSN 0009-5915

Olson, M.O. (2004). *The nucleolus*. Kluwer Academic Plenum Publisher, ISBN 0306478730, 9780306478734, London

Olson, M.O.; Hingorani, K. & Szebeni, A. (2002). Conventional and nonconventional roles of the nucleolus. *International Review of Cytology*, Vol. 219, pp. (199–266), ISSN 0074-7696

Oskarsson, T. & Trumpp, A. (2005). The Myc trilogy: Lord of RNA polymerases. *Nature Cell Biology*, Vol. 7, No. 3, pp. (215-217), ISSN 1465-7392

Pederson, T. (1998). The plurifunctional nucleolus. *Nucleic Acids Research*, Vol. 26, No. 17, pp. (3871–3876), ISSN 0305-1048

Penkina, M.V.; Karpowa, O.I. & Bogdanov, F.Y. (2002). Synaptonemal Complex Proteins: Specific Proteins of Meiotic Chromosomes. Molekuliarnaia Biologiia, Vol. 36, No. 3, pp. (397-407), ISSN 0026-8984

Pigozzi, M.I. & Solari, A.J. (1999). The ZW pairs of two paleognath birds from two orders show transitional stages of sex chromosome differentiation. *Chromosome Research*, Vol. 7, No. 7, pp. (541-551), ISSN 0967-3849

Pollock, D.L. & Fechheimer, N.S. (1978). The chromosomes of cockerels (*Gallus domesticus*) during meiosis. *Cytogenetics and Cell Genetics*, Vol. 21, No. 5, pp. (267-281), ISSN 0301-0171

Prieto, J.L. & McStay, B. (2005). Nucleolar biogenesis: the first small steps. Biochemical Society Transactions, Vol. 33, No. 6, pp. (1441-1443), ISSN 0300-5127

Rahn, M.I. & Solari, A.J. (1986). Recombination nodules in the oocytes of the chicken Gallus domesticus. *Cytogenetics and Cell Genetics*, Vol. 43, No. 3-4, pp. (187-193), ISSN 0301-0171

Raška, I.; Koberna, K.; Malinsk, J.; Fodlerova, H. & Masata, M. (2004). The nucleolus and transcription of ribosomal genes. *Biology of the Cell*, Vol. 96, No. 8, pp. (579–594), ISSN 0248-4900

Raška, I.; Shaw, P.J. & Cmarko, D. (2006). New Insights into Nucleolar Architecture and Activity. *International Review of Cytology*, Vol. 255, pp. (177-235), ISSN 0074-7696

Rodionov, A.V. (1996). Micro vs. macro: structural-functional organization of avian micro- and macrochromosomes. Genetika, Vol. 32, No. 5, pp. (597-608), ISSN 0016-6758

Ruas, M. & Peters, G. (1998): The p16INK4a/CDKN2A tumor suppressor and its relatives. Biochimica et Biophysica Acta, Vol. 1378, No. 2, pp. (F115-F177), ISSN 0006-3002

Ryan, K.M.; Philips, A.C. & Vousden, K.H. (2001). Regulation and function of the p53 tumor suppressor protein. *Current opinion in cell biology*, Vol. 13, No. 3, pp. (332–337), ISSN 0955-0674

Santoro, R. & Grummt, I. (2001). Molecular mechanisms mediating methylation dependent silencing of ribosomal gene transcription. *Molecular Cell*, Vol. 8, No. 3, pp. (719–725), ISSN 1097-2765

Santos, J.L.; del Cerro, A.L. & Díez, M. (1993): Spreading synaptonemal complexes from the grasshopper Chorthippus jacopsi: pachytene and zygotene obserwations. *Hereditas*, Vol. 118, No. 3, pp. (235-241), ISSN 0018-0661

Scheer, U. & Benavente, R. (1990). Functional and dynamic aspects of the mammalian nucleolus. *BioEssays*, Vol. 12, No. 1, pp. (14-21), ISSN 0265-9247

Scheer, U. & Hock, R. (1999). Structure and function of the nucleolus. *Current opinion in cell biology*, Vol. 11, No. 3, pp. (385–390), ISSN 0955-0674

Schmekel, K. & Daneholt, B. (1998). Evidence for close contact between recombination nodules and the central element of the synaptonemal complex. *Chromosome Research*, Vol. 6, No. 3, pp. (155-159), ISSN 0967-3849

Schmid, M.; Nanda, I.; Hoehn, H.; Schartl, M.; Haaf, T.; Buerstedde, J.M.; Arakawa, H., Caldwell, R.B.; Weigend, S.; Burt, D.W.; Smith, J.; Griffin, D.K.; Masabanda, J.S.; Groenen, M.A.M.; Crooijmans, R.P.M.A.; Vignal, A.; Fillon, V.; Morisson, M.; Pitel, F.; Vignoles, M.; Garrigues, A.; Gellin, J.; Rodionov, A.V.; Galkina, S.A.; Lukina, N.A.; Ben-Ari, G.; Blum, S.; Hillel, J.; Twito, T.; Lavi, U.; David, L.; Feldman, M.W.; Delany, M.E.; Conley, C.A.; Fowler, V.M.; Hedges, S.B.; Godbout, R.; Katyal, S.; Smith, C.; Hudson, Q.; Sinclair, A. & Mizuno, S. (2005). Second Report on Chicken Genes and Chromosomes 2005. *Cytogenetics and Genome Research*, Vol. 109, No. 4, pp. (415-479), ISSN 1424-8581

Shaw, P.J. & Jordan, E.G. (1995). The nucleolus. *Annual Review of Cell and Developmental Biology*, Vol. 11, pp. (93–121), ISSN 1081-0706

Smetana, K.; Klamova, H.; Pluskalova, M.; Stockbauer, P.; Jiraskova, I. & Hrkal, Z. (2005). Intranucleolar translocation of AgNORs in early granulocytic precursors in chronic myeloid leukaemia and K 562 cells. *Folia Biologica (Praha)*, Vol. 51, No. 4, pp. (89-92), ISSN 0015-5500

Smetana, K.; Klamova, H.; Pluskalova, M.; Stockbauer, P. & Hrkal, Z. (2006). To the intranucleolar translocation of AgNORs in leukemic early granulocytic and plasmacytic precursors. *Histochemistry and Cell Biology*, Vol. 125, No. 1-2, pp. (165-170), ISSN 0948-6143

Smith, J.; Paton, I.R.; Bruley, C.K.; Windsor, D.; Burke, D.; Ponce de Leon, F.A. & Burt, D.W. (2000). Integration of the genetic and physical maps of the chicken macrochromosomes. *Animal Genetics*, Vol. 31, No. 1, pp. (20-27), ISSN 0268-9146

Solari, A.J. (1977). Ultrastructure of the synaptic autosomes and ZW biwalent in chicken oocytes. *Chromosoma*, Vol. 64, No. 2, pp. (155-165), ISSN 0009-5915

Solari, A.J.; Fechheimer, N.S. & Bitgood, J.J. (1988). Pairing of ZW gonosomes and the localized recombination nodule in two Z-autosome translocation in *Gallus domesticus*. *Cytogenetics and Cell Genetics*, Vol. 48, No. 3, pp. (130-136), ISSN 0301-0171

Solovei, I.V; Gaginskya, E.R. & Macgregor, H.C. (1994). The arrangement and transcription of telomere DNA seqences at the ends of lampbrush chromosomes of birds. *Chromosome Research*, Vol. 2, No. 6, pp. (460-470), ISSN 0967-3849

Trere, D.; Ceccarelli, C.; Montanaro, L.; Tosti, E. & Derenzini, M. (2004). Nucleolar size and activity are related to pRb and p53 status in human breast cancer. *The journal of histochemistry and cytochemistry*, Vol. 52, No. 12, pp. (1601–1607), ISSN 0022-1554

Turner, P.C.; McLennan, A.G.; Bates, A.D. & White, M.R.H. (2004). *Biologia molekularna* (two edition), PWN, ISBN: 978-83-01-14146-2, Warsaw, Poland.

Wong, J.M.; Kusdra, L. & Collins, K. (2002). Subnuclear shuttling of human telomerase induced by transformation and DNA damage. *Nature Cell Biology*, Vol. 4, No. 9, pp. (731–736), ISSN 1465-7392

Permissions

The contributors of this book come from diverse backgrounds, making this book a truly international effort. This book will bring forth new frontiers with its revolutionizing research information and detailed analysis of the nascent developments around the world.

We would like to thank Dr. Andrew Swan, for lending his expertise to make the book truly unique. He has played a crucial role in the development of this book. Without his invaluable contribution this book wouldn't have been possible. He has made vital efforts to compile up to date information on the varied aspects of this subject to make this book a valuable addition to the collection of many professionals and students.

This book was conceptualized with the vision of imparting up-to-date information and advanced data in this field. To ensure the same, a matchless editorial board was set up. Every individual on the board went through rigorous rounds of assessment to prove their worth. After which they invested a large part of their time researching and compiling the most relevant data for our readers. Conferences and sessions were held from time to time between the editorial board and the contributing authors to present the data in the most comprehensible form. The editorial team has worked tirelessly to provide valuable and valid information to help people across the globe.

Every chapter published in this book has been scrutinized by our experts. Their significance has been extensively debated. The topics covered herein carry significant findings which will fuel the growth of the discipline. They may even be implemented as practical applications or may be referred to as a beginning point for another development. Chapters in this book were first published by InTech; hereby published with permission under the Creative Commons Attribution License or equivalent.

The editorial board has been involved in producing this book since its inception. They have spent rigorous hours researching and exploring the diverse topics which have resulted in the successful publishing of this book. They have passed on their knowledge of decades through this book. To expedite this challenging task, the publisher supported the team at every step. A small team of assistant editors was also appointed to further simplify the editing procedure and attain best results for the readers.

Our editorial team has been hand-picked from every corner of the world. Their multi-ethnicity adds dynamic inputs to the discussions which result in innovative outcomes. These outcomes are then further discussed with the researchers and contributors who give their valuable feedback and opinion regarding the same. The feedback is then collaborated with the researches and they are edited in a comprehensive manner to aid the understanding of the subject.

Apart from the editorial board, the designing team has also invested a significant amount of their time in understanding the subject and creating the most relevant covers. They scrutinized every image to scout for the most suitable representation of the subject and create an appropriate cover for the book.

The publishing team has been involved in this book since its early stages. They were actively engaged in every process, be it collecting the data, connecting with the contributors or procuring relevant information. The team has been an ardent support to the editorial, designing and production team. Their endless efforts to recruit the best for this project, has resulted in the accomplishment of this book. They are a veteran in the field of academics and their pool of knowledge is as vast as their experience in printing. Their expertise and guidance has proved useful at every step. Their uncompromising quality standards have made this book an exceptional effort. Their encouragement from time to time has been an inspiration for everyone.

The publisher and the editorial board hope that this book will prove to be a valuable piece of knowledge for researchers, students, practitioners and scholars across the globe.

List of Contributors

Karishma Collette and Györgyi Csankovszki
Department of Molecular, Cellular and Developmental Biology, University of Michigan, USA

Abrahan Hernández-Hernández
Department of Cell and Molecular Biology, Karolinska Institute, Stockholm, Sweden

Rosario Ortiz Hernádez and Gerardo H. Vázquez-Nin
Laboratory of Electron Microscopy, Faculty of Sciences, National Autonomous University of Mexico, Mexico

Agnieszka Kiełkowska
University of Agriculture in Krakow, Poland

Adela Calvente and José L. Barbero
Cell Proliferation and Development Department, Centro de Investigaciones Biológicas, Spain

Douglas Araujo
Universidade Estadual de Mato Grosso do Sul-UEMS, Unidade Universitária de Ivinhema, Brazil

Marielle Cristina Schneider
Universidade Federal de São Paulo-UNIFESP, Campus Diadema, Brazil

Emygdio Paula-Neto and Doralice Maria Cella
Universidade Estadual Paulista-UNESP, Campus Rio Claro, Brazil

Yoshihiro H. Inoue, Chie Miyauchi, Tubasa Ogata and Daishi Kitazawa
Insect Biomedical Research Center, Kyoto Institute of Technology, Japan

Aroldo Cisneros and Noemi Tel-Zur
French Associates Institute for Agriculture and Biotechnology of Drylands, The Jacob Blaustein Institutes for Desert Research (BIDR), Ben-Gurion University of the Negev (BGU), Beer Sheva, Israel

Chizue Hiruta
Department of Natural History Sciences, Graduate School of Science, Hokkaido University, Japan

Shin Tochinai
Department of Natural History Sciences, Graduate School of Science, Hokkaido University, Japan
Department of Natural History Sciences, Faculty of Science, Hokkaido University, Japan

Katarzyna Andraszek and Elżbieta Smalec
Siedlce University of Natural Sciences and Humanities, Poland

Printed in the USA
CPSIA information can be obtained
at www.ICGtesting.com
JSHW011359221024
72173JS00003B/342